TRANSPECIES DESIGN

In May 2019, the United Nations released the *Global Assessment Report on Biodiversity and Ecosystem Services* which warned that human activities will drive nearly one million species to extinction in a few decades. The primary reasons for this are habitat loss and biodiversity demise caused by changing climate, pollution, introducing nonindigenous species, clearing land, over population, and consumption. Given this situation, humans must change course as both human wellbeing and the wellbeing of other-than-human species are imbricated in one another. One way humanity can accomplish the needed transformation is to move beyond an anthropocentric view of life by embracing a transpecies approach that is premised upon interconnected flourishing.

Transpecies design, as outlined in this book, offers a new approach to regenerating the natural environment while honoring biodiversity. Rather than presenting the human experience as the goal of design, transpecies design takes the inextricable linkages connecting living things as both its starting point and end goal. As such, it moves beyond human experience serving as the fundamental ingredient for making better design processes and decisions.

This book is essential reading for artists, designers, and architects, as well as students of architecture, landscape architecture, interior architecture, art, product design, urban design, planning, environmental philosophy, and cultural studies.

Adrian Parr Zaretsky is the Dean of the College of Design at the University of Oregon, UNESCO Chair of Water and Human Settlements, and a Senior Fellow of the Design Futures Council. She is a philosopher, storyteller, and creative practitioner. She curated the Transpecies Design exhibition for the European Cultural Center's *Space, Time, Existence* 2023 Venice Architecture Biennale and the Watershed Urbanism exhibition for the 2021 ECC Venice Architecture Biennale. Her documentaries and art films have received numerous awards at independent film festivals around the world. She has published extensively on environmental culture and politics and her most recent book publication *Earthlings* (Columbia University Press 2022) earned a silver medal at the 2023 Nautilus Books Awards (environment category). Other publications include the trilogy: *Birth of a New Earth* (Columbia University Press 2017), *The Wrath of Capital* (Columbia University Press 2012), and *Hijacking Sustainability*

(MIT Press, 2009). She is the editor, with Santiago Zabala, of the *Outspoken* series published with McGill University Press.

Michael Zaretsky, AIA, is an Associate Professor and Head of the Department of Architecture in the College of Design at the University of Oregon. Zaretsky is a licensed architect with extensive experience in local and international design/build projects. His research is focused around culturally and environmentally responsive public interest design projects and community engagement with underserved communities locally and internationally. His published work includes *Precedents in Zero-Energy Design: Architecture and Passive Design in the 2007 Solar Decathlon* (Routledge Press, 2009) and *New Directions in Sustainable Design*, with Dr. Adrian Parr Zaretsky (Routledge Press, 2010). Zaretsky has articles published in several architectural journals and has presented at conferences around the world on Sustainability, Humanitarian Design, Public Interest Design, Design/Build and Community Engagement. From 2008–18, he was the Director of Design for the Roche Health Center in rural Tanzania, a Village Life Outreach Project. Roche Health Center is the first-ever permanent healthcare facility in this region. The Roche Health Center opened on April 1, 2011, and provides health care to as many as 20,000 villagers. From 2011–18, he was the Director of MetroLAB Design/Build, an academic community design/build program at the University of Cincinnati. His work is included in the 2018 book *The Public Interest Design Education Guidebook*, edited by Bell and Abendroth.

TRANSPECIES DESIGN

Design for a Posthumanist World

Edited by Adrian Parr Zaretsky and Michael Zaretsky

LONDON AND NEW YORK

Designed cover image: © Getty Images

First published 2025
by Routledge
4 Park Square, Milton Park, Abingdon, Oxon OX14 4RN

and by Routledge
605 Third Avenue, New York, NY 10158

Routledge is an imprint of the Taylor & Francis Group, an informa business

© 2025 selection and editorial matter, Adrian Parr Zaretsky and Michael Zaretsky; individual chapters, the contributors

The right of Adrian Parr Zaretsky and Michael Zaretsky to be identified as the authors of the editorial material, and of the authors for their individual chapters, has been asserted in accordance with sections 77 and 78 of the Copyright, Designs and Patents Act 1988.

All rights reserved. No part of this book may be reprinted or reproduced or utilised in any form or by any electronic, mechanical, or other means, now known or hereafter invented, including photocopying and recording, or in any information storage or retrieval system, without permission in writing from the publishers.

Trademark notice: Product or corporate names may be trademarks or registered trademarks, and are used only for identification and explanation without intent to infringe.

British Library Cataloguing-in-Publication Data
A catalogue record for this book is available from the British Library

ISBN: 978-1-032-51692-9 (hbk)
ISBN: 978-1-032-51689-9 (pbk)
ISBN: 978-1-003-40349-4 (ebk)

DOI: 10.4324/9781003403494

Typeset in Times New Roman
by KnowledgeWorks Global Ltd.

CONTENTS

Contributors *vii*

Introduction 1
Adrian Parr Zaretsky and Michael Zaretsky

SECTION I
Co-Habitation 7

1 How Does Ecological Science Support Transpecies Design? 9
 Steward Pickett

2 Interspecies, Multispecies, or Transpecies Design? What's the Difference? 16
 Adrian Parr Zaretsky

3 Transpecies Design and Biomimicry 24
 Henry Dicks

4 Design Against Extinction: Multispecies Methods and Engineered
 Living Materials 33
 Mitchell Joachim

5 Garden City to City in Nature: A Case for the Cohabitation of
 Tidal Ecologies along Singapore's Urban Waterfront 39
 Gabriel Tenaya Kaprielian

6 Floating-With: Buoyant Ecologies of Collaboration and Solidarity 51
 Adam Marcus, Margaret Ikeda, and Evan Jones

7 Unbecoming Human: Patricia Piccinini's Bioart and
   Postanthropocentric Posthumanism   56
   Kate Mondloch

8 Salt Formations   71
   Rosalea Monacella

SECTION II
Co-Creation   **79**

9 Origin and Evolution of Biodiversity: A Story of Life on Earth   81
   Christian Sardet

10 Adrian Parr Zaretsky in Conversation With Janet Laurence   90
   Adrian Parr Zaretsky and Janet Laurence

11 Birdsong and the Transpecies Aesthetic   93
   David Rothenberg

12 Entangled Intelligences: Transpecies Dialogues of Art   105
   Jiabao Li

13 Designing With Non-Humans: Ralph Ghoche in Conversation With
   Joyce Hwang   115
   Ralph Ghoche and Joyce Hwang

14 The Problem Is the Burning House   127
   Catherine Page Harris

15 Everything With Wings   136
   Sarah Walko and Gabriel Willow

16 Adrian Parr Zaretsky in Conversation With Carla Bengtson   144
   Adrian Parr Zaretsky and Carla Bengtson

17 Piñon Passage   150
   Nina Elder

Bibliography   *158*
Index   *166*

CONTRIBUTORS

Carla Bengtson is an artist who makes work by, for, and with other species in collaboration with scientists, dancers, composers, and perfumers. Bengtson has received awards from the NEA, the Ford Family Foundation, the Oregon Arts Commission, the Institute of Art and Olfaction, and an OAC Integrative Sciences Award. She has exhibited at the Cincinnati Contemporary Arts Center, the San Francisco Museum of Craft and Design, the Queens Museum, Craft Contemporary LA, and the Portland Art Museum. Artist residencies include Djerassi, Ucross, Mass MoCA, 18th St Arts Center, Signal Fire, and multiple research residencies at the Tiputini Biodiversity Station in the Amazon. She is a Professor in the Department of Art at the University of Oregon, and holds a BFA from Tyler School of Art, an MFA from Yale School of Art, and was a two-time participant in the Whitney Independent Study Program.

Henry Dicks works as an environmental philosopher at University Jean Moulin Lyon 3. A specialist in the philosophy of biomimicry, he is the author of *The Biomimicry Revolution: Learning from Nature how to Inhabit the Earth* (New York: Columbia University Press, 2023) and has published articles on this topic in a wide variety of journals, including *Philosophy of Science, Philosophy and Technology*, and *Environmental Ethics*. He is also a member of the UN Harmony with Nature Programme and co-pilots the Committee for Ethics and Environmental Responsibility at the CEEBIOS, France's leading biomimicry institute.

Nina Elder Artist and researcher Nina Elder creates projects that reveal humanity's dependence on and interruption of the natural world. With a focus on changing cultures and ecologies, Elder advocates for collaboration, fostering relationships between institutions, artists, scientists, and diverse communities. Her work takes many forms, including drawings, performance, pedagogy, critical writing, long-term community-based projects, and public art. Recent solo exhibitions of Elder's work have been organized by SITE Santa Fe, Indianapolis Contemporary, and university museums across the US. Her work has been featured in *Art in America*, *VICE Magazine*, and on PBS; her writing has been published in *American Scientist* and *Edge Effects Journal*. Nina's research has been supported by the Andy Warhol Foundation, the Rauschenberg Foundation, the Pollock Krasner Foundation, and the Mellon Foundation. Nina is an affiliate artist of the

National Performance Network. She has recently held research positions at the Center for Art + Environment at the Nevada Museum of Art, the Anchorage Museum, and the Art and Ecology Program at the University of New Mexico. Nina migrates between rural places in New Mexico and Alaska. https://www.ninaelder.com/

Catherine Harris is an Associate Professor and teaches Art and Ecology and Landscape Architecture at the University of New Mexico in a split position with the College of Fine Arts and the School of Architecture and Planning. She received her BA from Harvard University, 1988, MLA from UC Berkeley, 1997, and MFA from Stanford University, 2005. Harris works in art/design, and digital/analog expressions. Her built work resides at The Sevilleta National Wildlife Refuge, Socorro, NM, Marble House Project, Dorset, VT, Deep Springs College, White Mountains, CA, McCovey Field, SF, CA and The Violin Shop in Albuquerque, NM, among other sites. Recent projects include the sharing shelves, a creative placemaking project, the Red Water Pond Road Community Peace Center, an NEA ArtWorks funded project to create a space to memorialize the impacts of legacy mining on indigenous land, and sharing a drink, using video to create 3D drinking forms based on how animals drink water from trail cameras. Trans-species Repast—sharing meals with animals throughout northern Denmark and Vermont, USA—is an exploration of hierarchy, resources, and landscape. Trans-species Repast was shown at the UNM Art Museum (2016), the Land Shape Festival (2015) in Hanstholm, DK, Marble House Project, Dorset, VT (2015) and the Wignall Museum, CA (2014). Current research includes exploring indigenous landscape architecture, pursuing vertical definitions of urban wilderness in densifying cities, and posthuman places in the urban fabric. Pedagogical foci include indigeneity, video in landscape architecture, designing for climate change and introducing design skills into art practice.

Joyce Hwang is an Associate Professor and Director of Graduate Studies of Architecture at the University at Buffalo, State University of New York, and Founder of Ants of the Prairie. She is a recipient of the Exhibit Columbus University Research Design Fellowship (2020–21), the Architectural League Emerging Voices Award (2014), the New York Foundation for the Arts (NYFA) Fellowship (2013), the New York State Council on the Arts (NYSCA) Independent Project Grant (2013, 2008), and the MacDowell Fellowship (2016, 2011). Her work has been featured by the Museum of Modern Art (MoMA) and exhibited at Brooklyn Botanic Garden, Matadero Madrid, the Venice Architecture Biennale, and the Rotterdam International Architecture Biennale, among other venues. Hwang serves as a Core Organizer for Dark Matter U, and was previously on the editorial board for the *Journal of Architectural Education* (JAE). Hwang is a registered architect in New York State and has practiced professionally with offices in New York, Philadelphia, San Francisco, and Barcelona.

Margaret Ikeda is an Associate Professor at California College of the Arts, co-director of the Architectural Ecologies Lab, and a founding partner at ASSEMBLY, a practice whose focus, since 1994, is exploring connections, both in innovative details, materials and assemblies, and in the integration of multiple building specialists into the design process. Margaret brings an ability to form alliances with community stakeholders that builds a collaborative network for the realization of projects.

Mitchell Joachim, PhD, Assoc. AIA, is the co-founder of Terreform ONE and an Associate Professor of Practice at New York University. Formerly, he was an architect at the offices of Frank Gehry and I.M. Pei. He has been awarded a Fulbright Scholarship and fellowships with TED, Moshe Safdie, and Martin Society for Sustainability, MIT. He was chosen by *Wired* magazine for "The Smart List" and selected by *Rolling Stone* for "The 100 People Who Are Changing America." Mitchell won many honors, including: Lafarge Holcim Acknowledgement Award, Ove Arup Foundation Grant, ARCHITECT R+D Award, AIA New York Urban Design Merit Award, 1st Place International Architecture Award, Victor Papanek Social Design Award, Zumtobel Group Award for Sustainability, Architizer A+ Award, History Channel Infiniti Award for City of the Future, and *Time* magazine's Best Invention with MIT Smart Cities. He's featured as "The NOW 99" in *Dwell* magazine and "50 Under 50 Innovators of the 21st Century" by Images Publishers. He co-authored four books, *Design with Life: Biotech Architecture and Resilient Cities*, *XXL-XS: New Directions in Ecological Design*, *Super Cells: Building with Biology*, and *Global Design: Elsewhere Envisioned*. His design work has been exhibited at MoMA and the Venice Biennale. He earned PhD at Massachusetts Institute of Technology, MAUD at Harvard University, and M.Arch at Columbia University.

Evan Jones is an Adjunct Professor at California College of the Arts and co-director of the Architectural Ecologies Lab. He founded ASSEMBLY, a Berkeley-based architecture office, with Margaret Ikeda in 1994. ASSEMBLY's projects span in scale from furniture to multistory mixed-use housing. As implied by the name, the firm focuses on connections between physical materials and the collaborative process of design. Its projects have included the design and fabrication of installations and furnishings as well as large-scale commercial work.

Gabriel Kaprielian is an Assistant Professor of Architecture at the California Polytechnic State University San Luis Obispo and the Director of *Design and Innovation for Sustainable Cities* program at UC Berkeley. His research and creative work explore the interconnected relationship of the built and natural environment to reveal place-based narratives of past, present, and future that inform more resilient and livable cities. The focus of his current research is on the contested territory of urban waterfronts in tidal cities. This work seeks to reframe the problems associated with sea level rise and coastal climate change events as an opportunity to reshape the edge condition between land and sea, proposing new relationships between urban life and ecological systems. Kaprielian has served as a U.S. Fulbright Research Scholar in Singapore, American Arts Incubator Lead Artist for Peru, Fellow at the Exploratorium: The Museum of Science, Art and Humanities and Artist-in-Resident at Autodesk Pier 9 Workshop in San Francisco and Kunstnarhuset Messen in Norway.

Janet Laurence is a leading Sydney-based artist who exhibits nationally and internationally. Her practice examines our physical, cultural and conflicting relationship to the natural world. She creates immersive environments that navigate the interconnections between organic elements and systems of nature. Within the recognized threat of climate change she explores what it might mean to heal, the natural environment, fusing this with a sense of communal loss and search for connection with powerful life-forces. Her work is included in museum, university, corporate and private collections as well as within architectural and landscaped public places. Laurence has been a recipient of Rockefeller, Churchill, and Australia Council fellowships; recipient of the Alumni Award for Arts, UNSW; Hanse-Wissenschaftskolleg (HWK) fellowship; and Australian Antarctic Fellowship 2021 Artist in residence at the Australian Museum; she has been the

Australian representative for the *COP21/FIAC, Artists 4 Paris Climate* 2019, major solo survey exhibition at the MCA Sydney 2020, and at the Yu Hsui Museum of Art in Taiwan.

Jiabao Li creates works addressing climate change, interspecies co-creation, humane technology, and perceptions. Her mediums include wearable, robot, AR/VR, performance, scientific experiment, installation. In Jiabao's TED Talk, she uncovered how technology mediates the way we perceive reality. Jiabao is an Assistant Professor at The University of Texas at Austin and the founding director of the Ecocentric Future Lab. She is co-founder and chief product officer of Endless Health and previously of Snapi Health. In her four years at Apple, she invented and explored new technologies for future products, including the Apple Vision Pro. She graduated from Harvard GSD with Distinction and thesis award. Jiabao is the recipient of numerous awards, including Forbes China 30 Under 30, iF Design Award, Falling Walls, NEA, STARTS Prize, Fast Company, Core77, IDSA, A' Design Award, Webby Award, Cannes World Film Festival Award, and Outstanding Instructor Award. Her work has been exhibited internationally, at Venice Architecture Biennale, MoMA, Ars Electronica, Exploratorium, Today Art Museum Biennial, SIGGRAPH, Milan and Dubai Design Week, Ming Contemporary Museum, ISEA, Anchorage Museum, OCAT Contemporary Art Terminal, CHI, Museum of Design. Her academic papers have been published in top conferences and journals including SIGGRAPH, CHI, IEEE VIS, and Nature sub-journals. Her work has been featured on Fast Company, Art Forum, Business Insider, Bloomberg, Yahoo, South China Morning Post, TechCrunch, Domus, Yanko Design, Harvard Political Review, The National, Leonardo. https://www.jiabaoli.org/

Adam Marcus is an Associate Professor at Tulane University School of Architecture in New Orleans. He directs Variable Projects, a design and research studio that operates at the intersection of architecture, computation, and fabrication. He is also a partner in Futures North, a public art collaborative dedicated to exploring the aesthetics of data. His work explores ways in which new technologies can interface with craft, ornament, pattern, and ecological performance.

Rosalea Monacella, PhD, is an Academic at the Harvard University Graduate School of Design in the Landscape Architecture Program, an Architect and a registered Landscape Architect. Her research focuses on the techno-ecological indexing and reshaping of the city region through the notion of the "thickened ground" that describes the metabolism and material flows that generate the urban landscape vertically from the depths of the earth, through to atmospheric heights, and across an expanded ecology of political, economic, social, and environmental systems. Through this framework, speculations on alternative near-future landscapes explore innovative responses to climate change, shifting resource demands, and ecologies of energy.

Kate Mondloch is Professor of Contemporary Art and Theory in the Department of the History of Art and Architecture at the University of Oregon. She was the Founding Director of the New Media and Culture Certificate. She completed her undergraduate work in the School of Foreign Service at Georgetown University and earned both her MA and PhD in art history from UCLA. Mondloch's research interests focus on late 20th- and early 21st-century art, theory, and criticism, particularly as these areas of inquiry intersect with the cultural, social, and aesthetic possibilities of new technologies. Mondloch is the author of *Screens: Viewing Media Installation Art* (University of Minnesota Press, 2010) and *A Capsule Aesthetic: Feminist Materialisms in New Media Art* (University of Minnesota Press, 2018), for which she developed a related multimedia publication, *Installation Archive: A Capsule Aesthetic*, using the *Scalar* platform. She is currently working on a book tentatively titled *Art of Attention: Body-Mind Awareness and Contemporary Art*. Mondloch has been published in a variety of journals, including *Art Journal*, *Art Bulletin*, *Feminist Media Studies*, *Leonardo*, and *Vectors*, and has served on the editorial board of *Art Journal*, *Afterimage*, and *Media:Art:Write:Now*. She has been awarded research fellowships from the American Council of Learned Societies (ACLS), the University of California Humanities Research Institute, the Banff Centre, and the Oregon Humanities Center. Her research has also been supported by the Getty Research Institute and the Clark Art Institute.

Adrian Parr Zaretsky is the Dean of the College of Design at the University of Oregon, UNESCO Chair of Water and Human Settlements, and a Senior Fellow of the Design Futures Council. She is a philosopher, storyteller, and creative practitioner. She curated the Transpecies Design exhibition for the European Cultural Center's *Space, Time, Existence* 2023 Venice Architecture Biennale and the Watershed Urbanism exhibition for the 2021 ECC Venice Architecture Biennale. Her documentaries and art films have received numerous awards at independent film festivals around the world. She has published extensively on environmental culture and politics, and her most recent book publication *Earthlings* (Columbia UP, 2022) earned a silver medal at the 2023 Nautilus Books Awards (environment category). Other publications include the trilogy: *Birth of a New Earth* (Columbia 2017), *The Wrath of Capital* (Columbia 2012), and *Hijacking Sustainability* (MIT Press, 2009). She is the editor, with Santiago Zabala, of the *Outspoken* series published with McGill University Press.

Steward Pickett is an expert in the ecology of plants, landscapes, and urban ecosystems. Recipient of the Ecological Society of America's 2021 Eminent Ecologist Award, a member of the National Academy of Sciences, and the founding director of the Baltimore Ecosystem Study (1997–2016). Pickett's research focuses on the ecological structure of urban areas and vegetation dynamics, with national and global applications. Among his research sites: vacant lots in urban Baltimore, primary forests in western Pennsylvania, post-agricultural fields in New Jersey, China's rapidly urbanizing Yanqi Valley, and riparian woodlands and savannas in Kruger National Park, South Africa. By applying ecological theory to urban planning, architecture, and landscape architecture, Pickett strives to convert cities and suburbs from ecological liabilities into ecological assets. Using satellite data, Pickett studies urban landscape composition as it evolves and links this information to social and demographic influences.

David Rothenberg is a musician and philosopher. He wrote *Why Birds Sing*, *Bug Music*, *Survival of the Beautiful* and many other books, published in at least eleven languages. He has released more than thirty recordings, including *One Dark Night I Left My Silent House*, which was released by ECM, and most recently *In the Wake of Memories* and *Faultlines*. He has performed or recorded with Pauline Oliveros, Peter Gabriel, Ray Phiri, Suzanne Vega, Scanner, Elliott Sharp, Umru, Iva Bittová, and the Karnataka College of Percussion. *Nightingales in Berlin* is his latest book and film. Rothenberg is Distinguished Professor of Philosophy and Music at the New Jersey Institute of Technology.

Christian Sardet is a founder of the Laboratory of Cell Biology at the Marine Station of Villefranche-sur-Mer, IMEV, Centre National de la Recherche Scientifique (CNRS) and Sorbonne Université. Presently emeritus research director at the CNRS, Sardet is the author of numerous scientific publications on the cell and molecular biology of fertilization and development of embryos, and more recently on plankton. He is the recipient of the *Grand Prix des Sciences de la Mer* from the French Academy of Sciences. Creator of award-winning documentaries, animated films and DVDs, Christian received the European Award for Communication in Life Sciences from the European Molecular Biology Organization (EMBO). Christian Sardet was a co-founder and one of the scientific coordinators of the *Tara Oceans* expedition, devoted to a global study of plankton in all the oceans of the world. The "Plankton Chronicles" project was made to accompany the expedition. Initiated with CNRS Images (Véronique Kleiner, Catherine Balladur), the project continues today with Noé Sardet and Sharif Mirshak (Parafilms).

Sarah Walko is an artist, director, curator, and writer. She has her Master of Fine Arts degree from Savannah College of Art and Design and Bachelor of Arts from the University of Maryland. She is currently the Director of Visual and Performing Arts at Lenox Hill Neighborhood House and has directed non-profit arts organizations for fifteen years. Her visual art exhibitions have included: Raising the Temperature at the Queens Museum of Art, Preternatural at The Museum of Nature in Canada, Codex Dynamic Film Exhibition on the Manhattan Bridge Anchorage, Transcendence at Local Project in New York, Case Studies at Index Art Center in New Jersey, Baker's Dozen: 13 Artists on Found Objects at One Black Whisker Gallery in Pennsylvania, Fair Play, a group video exhibition in Miami, Florida, So That I Might Speak to You of Your Magnificence, a solo exhibition at The Teaching Gallery in New York, Rewoven, Innovation Fiber Arts Exhibition at the Queens Community College, CUNY in New York, I Embody at 310 Gallery, Marrietta College in Marrietta, Ohio, and Earth Revisited, a video exhibition on the Manhattan Bridge Anchorage in New York. She was an invited artist in the inaugural The First Ten, New Hope Artist Residency Program and selected as an inaugural participant in Art For Good: HATCHING A Better World program in 2020. She has been an artist in residence at many residency programs including Chateau Orquevaux, IPark, and the Elizabeth Foundation. She has been a visiting artist at Endicott College, Hudson Valley Community College, Kansas City Art Institute, University of Missouri, Roger Williams University, and Savannah College of Art and Design. She is a continuing NYFA Immigrant Artists Mentor within the mentorship program and a published author of fiction and nonfiction essays. She is a contributing writer in three anthologies that were published in 2022: *Sacred Promise* (Women Changing the World Press), *Neon Guides Me* (Praun & Guermouche) and *Royal Beauty* (Arts by the People).

Gabriel Willow is an artist and naturalist with a particular passion for birds. He explores the confluence of the human and natural environment in urban settings and shares these places with the public through lectures, tours, and artistic collaborations. He has collaborated with artists such as Mary Miss, Nina Katchadourian, George Trakas, and others exploring urban design and natural history. He also is an illustrator and DJ.

Michael Zaretsky, AIA, is an Associate Professor and Head of the Department of Architecture in the College of Design at the University of Oregon. Zaretsky is a licensed architect with extensive experience in local and international design/build projects. His research is focused around culturally and environmentally responsive public interest design projects and community engagement with underserved communities locally and internationally. His published work includes *Precedents in Zero-Energy Design: Architecture and Passive Design in the 2007 Solar Decathlon* (Routledge Press, 2009) and *New Directions in Sustainable Design*, with Dr. Adrian Parr (Routledge Press, 2010). Zaretsky has articles published in several architectural journals and has presented at conferences around the world on Sustainability, Humanitarian Design, and Public Interest Design. His work is included in the 2018 book *The Public Interest Design Education Guidebook edited* by Bell and Abendroth. He is the Director of Design for the Roche Health Center in rural Tanzania, a project of Village Life Outreach Project. Roche Health Center is the first-ever permanent healthcare facility in this region.

INTRODUCTION

Adrian Parr Zaretsky and Michael Zaretsky

The rapacious appetite of *Homo sapiens* to consume the earth's limited ecological goods and services is compromising other-than-human species' well-being and sustenance. In particular it is anthropogenic land and water use, along with the international nature of trade and tourism that has become a primary driver of global invasive species distribution, which are accelerating the rate of species extinctions.[1] Since life first began on earth there have been five mass extinctions of the earth's biota, all the result of natural events, such as an asteroid crashing into the earth, changes in oxygen levels, volcanic eruption, or rapid climatic heating or cooling. Whilst extinctions are a natural part of life on earth, today it is human activities which are driving the earth toward the brink of another mass extinction.[2] Currently, the number of species disappearing is both higher and faster than the baseline rate of one species per one million species a year. A study published in *Science* in 2014 shows that the rate and number of species going extinct is between 1,000 and 10,000 times faster as compared to a rate of 0.1 per million years during pre-human times.[3] What does this mean in concrete terms? First, the diversity between and within ecosystems and species is in decline. Second, the decline of apex species (at the top of the food chain) has 'far reaching effects on processes as diverse as the dynamics of disease; fire; carbon sequestration; invasive species; and biogeochemical exchanges among Earth's soil, water, and atmosphere.'[4]

It is not only other-than-human species that are at risk; humanity faces a significant existential threat the more damage it inflicts on the biosphere. The earth's ability to provide for human needs is exceeding the biosphere's ability to both produce sufficient ecosystem resources to meet that demand, as well as 'exceeding the regenerative and absorptive capacity of the biosphere.'[5] Human beings turn to nature for medicines, energy, food, and water. We rely on other-than-human species, such as bees, to pollinate plants that produce the fruits and vegetables we eat. The many healing properties of plants are an important source of medicine for nearly 4 billion people worldwide.[6] The rise in vector-borne and zoonotic diseases from 1990 to 2016 is connected to deforestation rates.[7] Meanwhile, the cultural identities of diverse human communities are often imbricated in rituals, beliefs, and values that involve historical and place-specific relationships formed with the natural world. For instance, visual artists seek inspiration from nature; shamans believe nature facilitates communication with the world of spirits; and

DOI: 10.4324/9781003403494-1

Hindus bathe in the Ganges River to purify themselves of sin. Put simply, human physical and emotional wellbeing is intertwined with biodiversity, which in turn is reliant upon clean air and water, soil quality, and relatively stable climatic and hydrological cycles. Put simply, human flourishing is existentially embedded in and intertwined with the flourishing of millions of other-than-human species.

It is critical that humanity does not complacently sit by as thousands of species move onto the lists of endangerment or extinction. Under the current circumstances, indifference and the refusal to change course are tantamount to one and the same thing: a speciesist slumber impeding the universal truth of existence from appearing. The truth in question is that a singular existence is constituted by a web of difference. A singular life is necessarily immersed in and realized through multiplicity and as such is conditioned by difference. In this context, transpecies design is emancipatory for it sets out to liberate *Homo sapiens* from the formal singularity of an individual or speciesist freedom by activating the collective agency that comes with biodiverse systems of mutual flourishing. The universality of this undertaking is that it rests in a common struggle in the sense that humans and other-than-humans alike will all be on the verge of extinction in the absence of biodiversity.

Transpecies design is a concept coined by Adrian Parr Zaretsky for an exhibition she curated as part of the 2023 European Cultural Center's *Space, Time, and Existence* exhibits held in conjunction with the Venice Architecture Biennale. Transpeciesism forms one theoretical pillar of Parr's tripartite theory of transenvironmentalism, as outlined in her 2022 publication, *Earthlings: Imaginative Encounters with the Natural World*. The other two theoretical pillars being transinternational and transgenerational thinking and practices. As this anthology sets out to demonstrate and articulate, transpecies design thinking and practices constitute a collective call to move beyond the anthropocentric basis of architectural and artistic practices to embrace design practices that are premised upon mutual flourishing for all of earth's species. In this regard, it picks up upon and deepens Jody Emel and Jennifer Wolch's call for transpecies urban theory, which calls for greater consideration of animal life in the design and planning of our cities by infusing a focus on biodiversity and mutual flourishing across a range of plant and animal lives.[8]

In this volume of essays we will begin our transpeciesist journey with an introduction to ecological science by world renowned ecologist Steward Pickett, who outlines the important role ecological science plays for the great decentering of *Homo sapiens*. Together Sardet and Pickett provide the reader with the fundamental scientific scaffolding for how a transpecies design practice might work.

Moving deeper into the philosophical and conceptual realm, Adrian Parr Zaretsky outlines how transpecies design refers to practices and thinking that seek to overcome, what Donna Haraway has described as 'human exceptionalism and bounded individualism.'[9] Teasing out the design implications associated with the use of different prefixes – *intra*species design, *multi*species design, *inter*species design, and *trans*species design – Parr Zaretsky describes how the use of the suffix 'intra' in design refers to a speciesist approach to design thinking and practices. Meanwhile, 'multi' species design indicates an additive design approach that contrasts multiple species experiences. 'Inter' species design focuses on species interaction as the point of departure for design practices. In this context, 'trans' species design moves beyond the human in design through a regenerative, restorative, and reconciliatory design project with practices of co-habitation and co-creation across species.

Henry Dicks continues the philosophical interrogation of the term transpecies design by investigating the methodological, ontological, and ethical components that constitute it.

Methodologically, transpecies design recognizes that design is not the exclusive domain of humans. Dicks maintains there is tremendous potential for the field of design if it embraces the agency of other-than-human species as co-designers. He notes that 'while there is an important sense in which transpecies design reduces the ontological divide between humans and other species, this is not to say the divide can be eliminated entirely.'[10] The ethical dimension of transpecies design is summoned by the call to eschew instrumentalizing other-than-human lives and a stated commitment to maximize the mutual flourishing of all species. The challenge for Dicks lies in the multi-temporal and open-ended temporality conditioning the designed world and here he infuses principles of biomimicry into the practice of transpecies design arguing for an 'iterative interplay of two principles: nature as model and nature as measure.'[11]

Mitch Joachim, founder of TerreformONE, tackles the question of what role living materials and biological systems can play in creating built environments that facilitate more generous human-nature interactions. The architectural practice of TerreformONE is premised on the idea that 'biology is technology' and as such the projects of TerreformONE set out to direct 'nature's intelligence' toward realizing restorative built environments that benefit all living species.[12] Gabriel Kaprielian follows Joachim by taking up the question of where nature ends and technology begins. Examining the massive restructuring and expansion of Singapore's shoreline and the significant loss of ecological coastal habitats, Kaprielian challenges the reader to a thought experiment: what if we imagined a future Singapore that 'embraced an adaptable waterfront that blurs the distinction between land and sea'?[13] This would result in a form of watershed urbanism that welcomes 'architectural strategies that are designed to accommodate water.'[14]

Moving into the fluidscapes of the ocean, Adam Marcus, Margaret Ikeda, and Evan Jones's Buoyant Ecologies project 'understands the ocean as an environment of dynamic flows, constant change, and collaboration across species.' Premised upon a notion of mutualism, their experimental floating structure equally invites human and other-than-human species dwelling; in this way it is an architectural strategy for unbecoming human. Picking up on the conceptual potential of unbecoming human, Kate Mondloch's study of bioart examines transpeciesist practices in terms of their ethical potential. She describes the fleshy hybrid creatures of Australian artist Patricia Piccinini as a medium for transpecies encounters that summon forth an ethical moment of care for nonhuman others through an 'inexhaustible and potentially destabilizing difference.'[15] The first section on co-habitation concludes with Rosalea Monacella's poetic and sensuous exploration of Kati Thanda (Lake Eyre) in Australia. Monacella continues the theme of mutual flourishing by invoking the problem of representation in fully expressing the many scales and temporalities that shape and transform landscapes.

At this point the volume shifts to the spaces of co-creation commencing with an origin story: the story of life on earth. Christian Sardet leads the reader through the building blocks of life, followed by a quick trot through the history of life, traversing the evolution of the first bacteria and archaea and onto the emergence of more complex cells followed by geochemical events, then the rise of a diverse mixture of organisms, until eventually ecosystems were formed and which to this day provide the co-creative foundation stones for biodiversity. Sardet's chapter is followed by a conversation between Adrian Parr Zaretsky and Australian multimedia artist Janet Laurence, a pioneering ecological artist who draws on a multiplicity of life forms to produce evocative and deeply moving transpeciesist spaces and experiences. Together Parr and Laurence explore the empathic thread central to a transpecies aesthetic. Laurence describes how she engages with extinction in the form and content of her work and how the transformative potential of art comes from arousing empathy in the viewer.

David Rothenberg deepens the connection between empathy, feeling, and transpeciesist acts of co-creating music with other-than-human animals. Rothenberg's exploration of animal music involves spontaneous and improvisational tactics, common amongst jazz musicians, to connect with the sounds and music of the other-than-human world. Similar to Rothenberg, artist Jiabao Li also utilizes personal narrative to provide an overview of different ways that artists, including herself, are engaging transpecies co-creation in their work. Li shares a series of works that address a shared creative process with squids, octopi, mice, bats, and glaciers in her work and then provides insightful examples and analyses of other artists and designers who are engaging transpecies co-creation.

In a reprinted interview from November 12, 2021 entitled 'Designing with Non-Humans' Ralph Ghoche interviews architect Joyce Hwang. Hwang is the founder of *Ants of the Prairie*, an architectural firm that addresses ecological issues through creative means. Hwang discusses her perspective on 'architect as advocate' including the history of projects for animals as well as a series of works designed by Hwang in collaboration with students and project teams in which the animals are the clients, users and stakeholders, while humans become the audience. Hwang and Ghoche discuss the work her teams have developed for multiple species of birds, bees, and bats and how these fit into the scope of work for non-human species. Catherine Page Harris further pursues the challenge posed in creating commensal communities with other-than-human species by engaging with acts of unconditional hospitality. As an artist and landscape architect, Harris creates landscapes of coexistence that bring humans, bees, cows, and other animals together through the sharing of meals.

Artists Sarah Walko and Gabriel Willow share with the reader the importance of interstitial and liminal spaces for not only art production and appreciation, but also in decentering the human by moving the practice and reception of art beyond the frame of human exceptionalism. In conversation with Adrian Parr Zaretsky, artist Carla Bengtson describes how her work is shaped through a series of transpecies collaborations that are grounded in the rigors of scientific hypothesis and experimentation. Like Walko and Willow, she speaks of the important role liminality plays in prompting spaces and times for curiosity and wonder to emerge, qualities that she views as enabling intimate exchanges amongst different species. With her contribution, artist Nina Elder concludes this anthology with a creative writing made of personal reflection and observation, tracing and expanding her sensorial and emotional range through encounters with other-than-human species.

The ideas and principles underpinning the term transpecies design is indebted to contemporary approaches to ecological design, the ecological and biological sciences, feminist new materialist philosophy, post-humanist thinking, the science of bioimicry, and indigenous approaches to cultural production and environmental stewardship, all of which appear throughout this anthology. All in all, transpecies design thinking and practices addresses, experiments with, and celebrates the myriad ways in which the vitality of human built environments is entangled with the wellbeing of a variety of species. Built environments constitute and are in turn constituted by intricate webs of life that sustain a variety of life forms, and which maintain the diversity between and within species. In this way, transpecies design is an other-than-anthropocentric approach to regenerating, restoring, reinvigorating, and replenishing the natural environment. Transpecies design moves beyond human experience and behavior serving as the fundamental ingredient for how design decisions are made, and the direction design processes take. Such practices and design thinking engage the substantive realities of a multiplicity of species as both design content and form, harnessing the affective capacity of design to maximize deep flourishing. Ultimately, the kaleidoscopic range that transpecies design traverses constitutes an emancipatory and inclusive politics that incorporates the biodiverse agency that comes from embracing and activating a wide variety of lives on earth.

Notes

1. Invasive species disrupt and impede ecosystem functioning, diminish species diversity, and remove native foods that native wildlife rely upon to survive.
2. Elizabeth Colbert, *The Sixth Extinction: An Unnatural History* (New York: Picador, 2014).
3. S. L. Pimm, C. N. Jenkins, R. Abell, T. M. Brooks, J. L. Gittleman, L. N. Joppa, P. H. Raven, C. M. Roberts, J. O. Sexton, "The biodiversity of species and their rates of extinction, distribution, and protection," *Science* 344, no. 6187 (30 May, 2014): 1246752-1–1246752-10. This study uses both the fossil record and analytical computer modeling to calculate extinction rates of pre-human and contemporary times.
4. James A. Estes, John Terborgh, Justin S. Brashares, Mary E. Power, Joel Berger, William J. Bond, Stephen R. Carpenter, Timothy E. Essington, Robert D. Holt, Jeremy B. C. Jackson, Robert J. Marquis, Lauri Oksanen, Tarja Oksanen, Robert T. Paine, Ellen K. Pikitch, William J. Ripple, Stuart A. Sandin, Marten Scheffer, Thomas W. Schoener, Jonathan B. Shurin, Anthony R. E. Sinclair, Michael E. Soulé, Risto Virtanen, and David A. Wardle, "Trophic downgrading of planet earth," *Science* 33 (15 July, 2011): 301–306; quote appeared on 301.
5. Michael Borucke, David Moore, Gemma Cranston, Kyle Gracey, Katsunori Iha, Joy Larson, Elias Lazarus, Juan Carlos Morales, Mathis Wackernagel, and Alessandro Galli, "Accounting for demand and supply of the biosphere's regenerative capacity: The National Footprint Accounts' underlying methodology and framework," *Ecological Indicators* 24 (2013): 519.
6. Intergovernmental Science-Policy Platform on Biodiversity and Ecosystem Services (IPBES) (2019), *The Global Assessment Report on Biodiversity and Ecosystem Services: Summary for Policymakers*, Bonn: 10. Accessed January 25, 2023. https://zenodo.org/record/3553579#.Y8grYOzMJsY
7. Serge Morand and Claire Lajaunie, "Outbreaks of vector-borne and zoonotic diseases are associated with changes in forest cover and oil palm expansion at global scale," *Frontiers in Veterinary Science* 8 (24 March, 2021). Accessed January 8, 2023. https://www.frontiersin.org/articles/10.3389/fvets.2021.661063/full
8. Jody Emel and Jennifer Wolch, *Animal Geographies: Place, Politics, and Identity in the Nature-Culture Borderlands* (London: Verso, 1998).
9. Donna Haraway, *Staying With The Trouble: Making Kin in the Chthulucene* (London: Duke University Press, 2016), 30.
10. See p. 27, this volume.
11. See p. 31, this volume.
12. See p. 35, this volume.
13. See p. 48, this volume.
14. See p. 48, this volume.
15. See p. 57, this volume.

SECTION I
Co-Habitation

1

HOW DOES ECOLOGICAL SCIENCE SUPPORT TRANSPECIES DESIGN?

Steward Pickett

Transpecies design calls for a great decentering.[1] This chapter examines the question, "Can ecological science help in the decentering toward transpecies design?" The best way to answer this question is to examine the decenterings that ecology itself is involved in. Some of these have a deep history, while others are only now emerging. I employ the viewpoint of one trained in the discipline of ecology, and so my first job is to explain the professional lens through which I might consider transpecies design. I then place ecological science in several key historical decenterings it has experienced or facilitated. These decenterings may also be important for transpecies design. They range across controversies about centering introduced into Western culture by the enlightenment. One of these decenterings helped give shape to contemporary science, but it simultaneously separated humans from nature conceptually. Although ecology was established in the late 19th and early 20th centuries, it has come to be useful in correcting the rift of people versus nature. Science itself is undergoing a decentering in response to its various social contexts as well. I will exemplify how ecology is currently being decentered through a multifaceted concept of coproduction, which may be a useful tool for promoting transpecies design.[2]

Ecology as a term can stand for three things: a body of knowledge; an approach to obtaining knowledge; and a metaphor or set of images to connect the first two to various social applications.[3] I start out with the ecologist's view of the discipline as a body of knowledge, knowing that these days, this may not be what ecology means to most people.

The Western science of ecology traces its roots to the European colonialist voyages of discovery, which revealed to that continent's elites an amazing variety of organisms, landscapes, and peoples. Western natural science emerged to name and catalog this richness, and explain the regularities and contrasts within it. Von Humboldt (1769–1859) established biogeography, Darwin (1809–1882) articulated the first confirmed theory of evolution based on a testable mechanism, and Lyell (1797–1875) and other geologists theorized a dynamic Earth, just to point to three "taproots" of contemporary ecological science.

The contemporary science of ecology can be defined like this: The scientific study of the processes influencing the distribution and abundance of organisms, the interactions among organisms, and the interactions between organisms and the transformation and flux of energy, matter, and information.[4] Information is a key component to this definition. Information can be in the minds of migrating individuals, encoded in the genetics of seeds, or expressed as cultural norms, news, art, capital, or credit, among other forms. These examples of information reinforce the idea, latent in the definition, that ecological systems contain humans, human artifacts, and human actions. These human components aren't just the result of individual behaviors, but they also emerge from interacting groups and networks: institutions and economies, households and governments, technologies, gangs, religious establishments, and social media.[5]

Ecology examines patterns of the various things mentioned in its definition, as well as how those patterns change through time or differ from place to place. But ecology is not content with merely describing patterns, or even relating different static patterns to each other. Fundamentally, ecology is concerned with why and how patterns come to exist, and what causes them to change. This is summarized in the call for understanding transformation in the definition of ecology. Understanding transformation requires the science to simultaneously deal with capacity for and limits to change. Ecological science supports an important generalization about transformation: "No natural system grows without limit," but the biological need to create order in the face of thermodynamic inevitability often pushes up against limits.

Ecology in the popular mind is sometimes associated with the aphorism that "Everything is connected to everything else." This might sound reasonable given the brief discussion of the definition, above. Let me propose an alternative generalization that is much more appropriate to the way that ecological scientists actually work: "Everything is connected to *something* else." In other words, we must discover what things are consequentially connected to each other. It is our job to ferret out the consequential interactions in a network of transformation. Furthermore, we also have to understand what interactions may arise from a distance – teleconnection is the Greek-infused term for that. And we have to discover what interactions create legacies that persist through time, or whose outcomes only appear indirectly. An example of an indirect interaction appears when a predator suppresses the activity of one of a pair of competing prey species, leading the second competitor to expand its population. In other words, ecology is all about discovering which among the myriad possible interactions in a place or situation generate structure and change through time.

These approaches show that ecologists can select their focus. On one hand, study can focus on specific places, or on the other, on a specific interaction. Focusing on a place requires setting or accepting boundaries, often those established for some non-ecological purpose, like a city or a farm. Within those chosen boundaries, research is designed to reveal what ecologically active things are present, and what connections there may be among them. For example, the flows and transformation of different chemical forms of nitrogen may be of interest. Some forms of nitrogen serve as limiting nutrients for the growth of plants, which cascades through an entire biotic ecosystem including animals and microbes as well as other plants. In lakes or streams, NO_3- can be a pollutant, and can be toxic to people when it contaminates their drinking water. This is a biogeochemical approach to the ecology of a place. Sometimes such places are called "systems" even before a detailed model of the fluxes and transformations in the place of interest are known. Ecologists are sometimes sloppy in using the word "system" to stand for different things: the technical model of the network and its transformations in a place, and for the place itself. Such places can be small sites or the biosphere of the entire Earth.

An alternative approach is to start with an interaction of interest. For example, if we are interested in the relationship between coyotes and urban pets, like small dogs or cats, the competitive or predatory behavior of the coyotes may be the starting focus. The first concern, then, would not be to bound a place as a system – except maybe quite broadly, like a whole city – but to start looking at coyote-pet relationships and see where that leads. Other components of the model might be outdoor feeding of pets, or letting pets roam outside the house. Furthermore, the large-scale territorial or foraging patterns of coyotes and their interaction with built or green infrastructure in a city might be of concern. This approach of focusing on some interaction of interest and seeing where and to what other interactions that interaction leads can be (impressively) called "progressive contextualization." Simply put, one builds a model of interactions by tracing consequences ever more broadly, sometimes widely in space, from the initial focus.

Now I move from ecology as a science to ecology as a metaphor or non-technical image. This distinction is important because in science the assumptions behind a model need to be clearly articulated, whereas metaphors tend to obscure assumptions.[6] The assumptions of models in science can be tested as fundamental hypotheses of the theory. Alternatively, model assumptions can be critically examined to explore whether they embody cultural values or social biases that might affect the structure or explanations offered by the theory. The exposure and evaluation of the role of bias and values is one of the most important outcomes of open communication within scientific community.[7] In contrast, when ecology is used as a metaphor, there is no guarantee that social and personal values and biases, desired social outcomes, or world views will not be in play. In fact, metaphorical uses often are intended to serve such values or points of view. Metaphors like "nature red in tooth and claw," "mature forest," "climax forest," or "resilience" can stimulate growth of theory, and point to things that need to be tested. Metaphors may also serve as images that translate scientific insights into civic discourse or the practice of design or environmental management.

So, as valuable as metaphors are, they can be trouble. For example, when they seem to express ecological ideas, but actually cloak social ideologies. Garrett Hardin's "lifeboat Earth" metaphor seems reasonable on its face, but hides his racist and anti-immigrant politics.[8] Even when metaphors don't disguise untoward assumptions, they may still cause confusion or misunderstanding at the intersection of science and society.

Metaphors may represent inappropriate analogies when moving from one discipline to another. For example, the ideas of plant succession were taken from ecology and applied to urban systems by the ground-breaking sociologists of the Chicago School as they worked to understand the massive changes in size and demography of their city in the early decades of the 20th century.[9] They explained the transformation of the city in terms of one human community replacing another, a strict arrangement of human communities into zones, and a fixed "life cycle of the city" reflecting replacement of one community by another, and ending, unfortunately in a blighted core. Although their misapplication of ecological theory was replaced in the 1930s, the idea of urban life cycles leading to blight persisted into the 1960s as a justification for urban renewal.[10] Note that "blight" is also a loaded metaphorical term.

Sometimes when a scientific theory advances or changes, an old, familiar metaphor may remain associated with the subject area. For example, "climax" referring to a terminal state of orderly vegetation dynamics is largely avoided now because the finality and maximal development it once implied are not core components of the theory.[11]

The contrast between ecology's founding Equilibrium paradigm and its contemporary, more inclusive non-equilibrium paradigm, shows the trap that metaphor can be.[12] Under the older set of background assumptions, mere mention of ecological systems concepts could imply that they were closed to material flows; self-regulating in their dynamics, undisturbed by external events, characterized by a single deterministic pathway of change; possessed of a stable end point; and lacking human influence or action.[13] The non-equilibrium paradigm is more inclusive, since it allows these earlier assumptions to govern special cases, but it also prompts ecologists and those who employ ecological knowledge to expect the opposite assumptions to often help understand systems in the world.

I have already mentioned that one of the legacies of the Western enlightenment was conceiving of nature and humans as separate, and indeed in opposition. On the other hand, the sciences that grew out of the enlightenment have served to decenter humans from their perch at the top of the "great chain of being." For example, the geocentric model of the world saw the Earth as the focus of the universe. Scientific observation and theory ultimately replaced that view with one placing the Sun at the center of our solar system, and ultimately conceived us as residing in one galaxy among multitudes.

The theory of evolution as an accepted explanation showed that the same processes that governed the changes across generations of plants and animals also pertained to humans. Later, the understanding of the genetic code confirmed this at a deep level of the machinery of inheritance and cellular metabolism.

The enlightenment also bequeathed us knowledge of the great variety and abundance of life on Earth. This was the observational raw material that was ultimately explained by evolutionary and genetic models. Studying both the structural similarities and differences in the diversity of life yielded solutions to many conundrums of the distribution of different types of organisms.

But the sciences of the enlightenment also provided traps for humanity beyond that of the human versus nature dichotomy. These are some of those additional traps:

- Biotic and geologic diversity were to be valued because of their utility to humanity, and as the fuel for progress in particular. The world was seen as a storehouse of resources for people to exploit.
- The exploitation was driven by colonialism, which appropriated land, labor, and sometimes people themselves to provide and work the resources of biological diversity. Thus access to the benefits of newly marshalled resources was not afforded to all.
- Colonialism was the parent of extractive capitalism, and of associated genocide that removed indigenous peoples from lands desired as sources of raw material by colonial powers.
- Racism was a seemingly biologically endorsed ranked differentiation among humans that justified enslavement, displacement, and genocide. In reality, racialization of people is a social tool created to serve colonial and within-nation exploitation of peoples.

Although these are serious traps, they are not inescapable. Just because a science is a child of colonialism, capitalism, and racism does not mean it must grow up to be colonialist, capitalist, or racist. Indeed, many scholars and activists are working to ensure that moving forward, ecological science is anti-racist.[14] If ecology can aspire to be anti-racist and anti-colonialist, might it not contribute to healing the human-nature divide that it also inherited from the enlightenment?

Decentering in the science of ecology may promote decentering humans in urban design. How might it do so? Within ecology, it is now common to recognize multiple pathways by which systems change – be they individual organisms, communities of plants or animals, or biogeochemical networks. For example, there are many detailed patterns of transition that may take place as vegetation changes after disturbance. In addition, the "endpoint" or relatively stable system may take many forms, depending on specific environmental conditions or historical legacies. Consequently, there can be a wide variety of ecological models to represent system structure and change. This recognition has even allowed ecological researchers to see their discipline as a plural endeavor. It is clearly legitimate to speak of "ecologies," as long as the nature of the approach or goal of a model is specified.[15] This plurality is in part also a result of the large number of specialist perspectives and social contexts which influence the knowledge that ecology can generate. But more fundamentally, the variety of plural ecologies results from a deep condition of the material world: The worlds we now inhabit are the product of both social processes and natural or biophysical processes.[16] That is, they are coproduced – their characteristics emerge from the intimate interaction of natural and human actions.

Such interactive nature-society entanglement means that society can call on ecology in all its plural glory. The variety of models of the coproduced world stands against the universality of modernism. Ecologies may stand against the bureaucratic state with its uniformity of policy and management, which ignore ecological contingency and probability. Ecologies may stand against powerful elites who require a unitary view of social values. This variety of calls on ecology suggests that the plural science can be a tool for social movements seeking to improve racial equity or environmental integrity.[17]

A decentered, plural science of ecology may serve the needs of decentered societies. Human diversity along many dimensions can employ the diverse, particularized models of ecology. Although human difference has been used as a lever to support colonialist, capitalist, and racist exploitation and oppression of peoples, it may be that ecologies perceived or modeled to represent the lived environments of different sexes, genders, economic classes, people of different national origins, or those experiencing contrasting immigration status can serve the cause of equity. A universal ecological model might not serve equity as effectively. Likewise different locations – zip codes, school districts, residential zones, topographic locations, proximity to hazards, for example – may best be represented by different models of ecological structure and process. In fact, global changes, such as the increasing frequencies and intensities of disasters originating from extreme events, require new theories or models that account for the new and dynamic interactions of sequences of events through time.[18] "Recurrent acute disasters" require new models – ecologies – of their effects increasingly linked through time. Ecologies in this sense may help expose vulnerabilities or capacities for recovery from hazard which can improve preparation or recovery from crises, be they of initially human or biogeophysical origin.[19] It will be important to decouple fate from difference, and use of pluralistic ecological models may facilitate that.

What Does Decentering People and Decentering in Ecology Mean for Transpecies Design?

Ecology in all its guises is about the structure and processes in biologically powered networks, those in which solar energy is the fundamental resource. (There are of course, a few ecosystems powered by "chemoautotrophs," organisms that use energy derived from breaking the bonds in some inorganic molecules, e.g., deep sea vents.) But energy isn't the only thing that flows in ecological networks. Various kinds of influence, either by direct interaction, or by flow of information, also characterize ecological networks. And these networks are not only regulated by internal interactions but also by interactions that are initiated beyond their borders. However, the globe seems now to be composed of interacting networks that are coproduced by the intimate interaction of humans – as both individuals and as social/political/economic/cultural institutions. Ecosystem services that provide resources, support the supply of resources, regulate the flows and conditions in ecosystems, or constitute cultural resources are all aspects of coproduction.[20] Global change in all its dimensions, including climate, urbanization, land change, resource depletion, transport of species, or toxification of the biosphere, have embedded human drivers.[21]

Design that takes decentering humans seriously will be design that can be sensitive to biodiversity, and can focus on design features necessary for human thriving but not necessarily for mere human convenience.[22] Displacing humans from the center of the design world can make space for the workings of nature – described as ecosystem services or nature-based solutions – that ultimately will favor humans, biodiversity, and ecosystem functioning. Decentering in ecology can help promote a diversity of models of ecological processes that can be valued or supported by a diverse array of human communities having different social identities and exposure to vulnerable situations. In other words, it is worth asking, can we take coproduction seriously enough to promote the decenterings required for transpecies design to transform a threatened world?

Notes

1 Jennifer R. Wolch, Jason Byrne, and Joshua P. Newell, "Urban Green Space, Public Health, and Environmental Justice," *Landscape and Urban Planning*, no. 125 (2014): 125.
2 Steward T. Pickett, Mary L. Cadenasso, and Anne M. Rademacher, "Toward Pluralizing Ecology: Finding Common Ground across Sociocultural and Scientific Perspectives," *Ecosphere* 13, no. 9 (2022), https://doi.org/10.1002/ecs2.4231.
3 Steward T. Pickett and Mary L. Cadenasso, "The Ecosystem as a Multidimensional Concept: Meaning, Model, and Metaphor," *Ecosystems* 5, no. 1 (2022): 5.
4 Kathleen C. Weathers, David L. Strayer, Gene E. Likens, eds., *Fundamentals of Ecosystem Science*, 2nd ed. (London: Academic Press, 2021).
5 William R. Burch, Jo Ellen Force, and Gary E. Machlis, *The Structure and Dynamics of Human Ecosystems: Toward a Model for Understanding and Action* (New Haven: Yale University Press, 2018).
6 Brendon Larson, *Metaphors for Environmental Sustainability: Redefining Our Relationship with Nature* (New Haven: Yale University Press, 2011); Steward T. Pickett, Mary L. Cadenasso, and J. M. Grove, "Resilient Cities: Meaning, Models, and Metaphor for Integrating the Ecological, Socio-Economic, and Planning Realms," *Landscape and Urban Planning* 69, no. 4 (2004).
7 Helen E. Longino, *Science and Social Knowledge: Values and Objectivity in Scientific Inquiry* (Princeton: Princeton University Press, 1990).
8 Eve Z. Bratman and William P. DeLince, "Dismantling White Supremacy in Environmental Studies and Sciences: An Argument for Anti-Racist and Decolonizing Pedagogies," *Journal of Environmental Studies and Sciences* 12, no. 2 (2022); Matto Mildenberger, "The Tragedy of the *Tragedy of the Commons*," *Scientific American* (blog), April 23, 2019, https://blogs.scientificamerican.com/voices/the-tragedy-of-the-tragedy-of-the-commons/.
9 Robert E. Park and Ernest W. Burgess, *The City* (Chicago: University of Chicago Press, 2019).
10 Jennifer S. Light, *The Nature of Cities: Ecological Visions and the American Urban Professions, 1920–1960* (Baltimore: Johns Hopkins University Press, 2009).
11 Scott J. Meiners, Steward T. Pickett, and Mary L. Cadenasso, *An Integrative Approach to Successional Dynamics: Tempo and Mode of Vegetation Change* (New York: Cambridge University Press, 2015).
12 Jianguo Wu and Orie L. Loucks, "From Balance of Nature to Hierarchical Patch Dynamics: A Paradigm Shift in Ecology," *The Quarterly Review of Biology* 70, no. 4 (1995).
13 Steward T. Pickett, V. Thomas Parker, and Peggy L. Fiedler, "The New Paradigm in Ecology: Implications for Conservation Biology Above the Species Level," in *Conservation Biology: The Theory and Practice of Nature Conservation, Preservation, and Management*, eds. Peggy Lee Fielder and Subodh K. Jain (New York: Chapman and Hall, 1992), https://doi.org/10.1007/978-1-4684-6426-9_4.
14 Pickett, Cadenasso, and Rademacher, "Toward Pluralizing Ecology;" Christopher J. Schell et al., "The Ecology and Evolutionary Consequences of Systematic Racism in Urban Environments," *Science* 369, no. 6510 (2020), https://doi.org/10.1126/science.aay4497.
15 Pickett, Cadenasso, and Rademacher, "Toward Pluralizing Ecology."

16 Anne Rademacher, Mary L. Cadenasso, and Steward T. Pickett, "From Feedbacks to Coproduction: Toward an Integrated Conceptual Framework for Urban Ecosystems," *Urban Ecosystems* 22, no. 1 (2019).
17 Leilani Nishime and Kim D. Hester Williams, eds., *Racial Ecologies* (Seattle: University of Washington Press, 2018).
18 Gary E. Machlis, Miguel O. Román, Steward T. Pickett, "A Framework for Research on Recurrent Acute Disasters," *Science Advances* 8, no. 10 (2022), https://doi.org/10.1126/sciadv.abk2458.
19 Mary L. Cadenasso, Anne M. Rademacher, Steward T. Pickett, "Systems in Flames: Dynamic Coproduction of Social-Ecological Processes," *BioScience* 72, no. 8 (2022).
20 Anke Fischer and Antonia Eastwood, "Coproduction of Ecosystem Services as Human-Nature Interactions—An Analytical Framework," *Land Use Policy* no. 52 (2016): 41–50.
21 Brian J. L. Berry, "Urbanization," in *The Earth as Transformed by Human Action: Global and Regional Changes in the Biosphere over the Past 300 Years*, B. L. Turner II et al., eds. (New York: Cambridge University Press, 1990); Peter M. Vitousek, Harold A. Mooney, Jane Lubchenco, and Jerry M. Mellilo, "Human Domination of Earth's Ecosystems," *Science*, no. 277 (1997). https://www.science.org/doi/10.1126/science.277.5325.494.
22 Timothy Beatley, *The Ecology of Place: Planning for Environment, Economy, and Community* (Washington, DC: Island Press, 1997).

2
INTERSPECIES, MULTISPECIES, OR TRANSPECIES DESIGN?

What's the Difference?

Adrian Parr Zaretsky

The earth is nearly 4.5 billion years old.[1,2] Approximately 150,000 years ago the species *Homo sapiens* populated what is today called East Africa. Just 80,000 years later *Homo sapiens* had gone on to become the dominant human species on earth.[3] Over the past couple of hundred years *Homo sapiens* have become the earth's 'dominant evolutionary force' dramatically changing the chemical makeup of the atmosphere and oceans, the climate, hydrological cycles, the number and diversity of species, and biodiversity.[4] By 1804 the human population on earth had reached 1 billion people and by 1927 it had doubled in size.[5] On November 15, 2022, the global human population reached 8 billion people.[6] During a short period of time in the earth's history *Homo sapiens* have overfished the seas; polluted air and water resources; genetically modified bacteria, animals, and plants; increased the average global temperature; and degraded or wiped out roughly two-thirds of the earth's original rainforest cover.[7] Anthropocentric activities present all life on earth, including *Homo sapiens*, with a quantitative and qualitative problem: the growing population and the environmental impact associated with how the increasing number of humans on earth live is pushing the biosphere toward a critical threshold and an irrevocable tipping point.

The earth's carrying capacity sits at approximately 7 billion people living at subsistence level.[8] To put this into context, July 28 was officially Earth Overshoot Day for the year 2022 – the day when the human consumption of ecological resources and services for that year exceeded the earth's capacity to sustainably replenish and meet that demand.[9] The Global Footprint Network uses UN data sets and additional information from peer-reviewed publications to create national footprint and biocapacity accounts. The six evaluation categories used to create the national footprint accounts are: carbon, fishing, croplands, the built environment, forest products, and land used for grazing. The accounts measure a nation's resource capacity and usage, and when taken together they measure the earth's annual carrying capacity to supply and meet human demand for ecosystem services. Two measurement systems are used to calculate carrying capacity. The first, ecological footprint, is 'a measure of the demand populations and activities place on the biosphere in a given year, given the prevailing technology and resource management of that year.'[10] The second, biocapacity, is a 'measure of the amount of biologically productive land and sea area available to provide the ecosystem services that humanity consumes – our ecological budget or nature's regenerative capacity.'[11] Currently, 85% of countries consistently run an ecological deficit, meaning their ecological footprint exceeds their biocapacity.[12]

In 1861 the human demand for ecosystem resources and services was estimated at 0.7 planets; by 2008 this had grown to 1.5 earths; by 2022 it reached 1.8 earths.[13] The organic carrying capacity of the earth is estimated to be around 2.4 billion people.[14] Meanwhile, it would take 5.1 earths if everyone lived like the average person did in the United States in 2022.[15] Unsurprisingly, as the Global Footprint Network has calculated, high income countries such as the United States overshoot significantly earlier each year than lower income countries. For 2023 the earth overshoot day for Jamaica was estimated to occur on December 20 of that year, whereas for the United States it was estimated to occur on March 13.[16] In addition, it is important to note that the earth overshoot calculations are not the organic carrying capacity of the earth, meaning a world without agriculture biotechnology.

Urbanization is one of the dominant forms that human-environmental interaction currently takes, with the majority of people on earth now living in urban areas. The World Bank estimates that by 2050 nearly 7 in 10 people will be living in urban areas.[17] Urbanization is transforming the earth across a variety of spatial and temporal scales, turning large areas of land into dark surfaces (roofs and roads that absorb as opposed to reflecting solar radiation), made of impermeable materials (asphalt, paving, concrete, glass, and steel) and carrying greater thermal mass from buildings. When taken together, these urban characteristics lead to the formation of urban heat islands, causing the annual mean temperature to be higher in these areas. This impacts health, air and water quality, prompting a vicious cycle for overall environmental quality as more people use more energy to cool their environment, which in turn increases global CO_2 emissions. In short, human population growth is spurring on the growth of 'anthromes,' a neologism coined by Earl Ellis and Navin Ramankutty in 2008 to reference the new anthropocentric biomes made up of human-built environments and agricultural land. Expanding the ecologist's definition of a biome, which describes 'global patterns of ecosystem form, process, and biodiversity,' Ellis and Ramankutty present an alternate ecological perspective of the terrestrial biosphere: anthropogenic biomes.[18] Their term 'anthromes' is one way to represent the many ways in which human activities have changed the earth's biodiversity and ecological processes. The term, anthrome, characterizes 'terrestrial biomes based on global patterns of sustained, direct human interaction with ecosystems,' and it acknowledges the extensive influence humans have had on global ecosystems with a view to 'moving us toward models and investigations of the terrestrial biosphere that integrate human and ecological systems.'[19] Human settlements, transportation and communication infrastructure, waste, and agricultural practices together make up the many ways that humans interact and transform the earth's ecosystems. Design is integral to all.

Inter, Multi, or Transpecies Design

Homo sapiens have created a gargantuan challenge for the viability of the majority of life forms on earth, including themselves. Despite having immense ingenuity, evolving ways of knowing and increasing technological sophistication, *Homo sapiens* are wrecking the very life support systems upon which they and countless other-than-human species depend. How do we shift course? How do we re-harness the ways in which we live in the world and the world in us? One way we can accomplish the needed transformation is to move beyond an anthropocentric view and embrace a transpecies one. More specifically, transpecies design practices and thinking could be harnessed to effectively respond to the wicked problem *Homo sapiens* have created for themselves and other-than-human species by transforming how built environments work. The idea is a simple one and it involves inviting into the design of the built environment diverse forms of inhabitation, such that built environments can function as habitats for a variety of species becoming a source of flourishing for both humans and other-than-humans alike.

An anthropocentric viewpoint is one that situates the human as the primary point of reference for how value is ascertained, produced, and articulated. As such it is a form of species discrimination, or what Australian ethicist Peter Singer might define as being 'speciesist.' Singer's definition of speciesism is that it constitutes a 'prejudice or attitude of bias in favor of the interests of members of one's own species and against those of members of other species.'[20] Anthropocentricism is therefore a moral bias whereby human beings regard human life as more valuable than other-than-human lives.[21] In this regard, anthropocentric design could be defined as intraspecies design. As the prefix *intra* signifies, intraspecies design would address environmental degradation from within the single viewpoint of how it impacts human beings. As such it presupposes a universal human subject as the primary point of reference for design decisions and ethical beneficiary of design goals and outcomes.

If innovation in the context of design comes from how it transforms people, the world in which people live, and ultimately in an age of rapid environmental degradation and climate change, how people live with each other and other-than-human neighbors and friends, then designers must inevitably engage in a meaningful way with this network of challenges. More specifically, how might design become more sensitive toward the needs of other-than-human species and address the rapid decline of other-than-human habitats? In posing the question this way the idea is to tease out the myriad ways in which language, imagery, cognitive maps, mediums of practice, and affects can form constellations of representation that either participate in the oppressive structures and systems of violence that humans have waged upon other-than-human species, or whether these are used to emancipate and pry open more generous and welcoming human perspectives, understandings, feelings, beliefs, and imaginaries toward other-than-human species. But why *transpecies* design? Why not multispecies or interspecies design? What is the difference? The different prefixes – *inter*, *multi*, and *trans* – point to some useful distinctions concerning how we think about, feel, know, and understand other-than-human relations through design, and all three prefixes describe different approaches to how species' relations work.

The prefix 'multi' in multispecies design suggests an approach to design that aspires toward interaction across species by adding in different species habitats as long as they don't disrupt human wellbeing and aesthetic sensibilities. Multispecies design in this regard does not break down the boundary between species as a goal or outcome of a given design for it entails minor interaction across species. It is a design practice premised upon contrasting one species against another. Multispecies design then refers to more formalist approaches to the natural world and design, for example, the ornamental use of natural patterns and shapes in façade structures, like the use of fractal geometries on the Federation Square facade in Melbourne, Australia.

Interspecies design combines the perspectives or experiences of two or more species, such that the speciesist boundaries establishing one species as separate to, or in opposition to, one another start to unravel. Leading with the prefix definition, the notion of interspecies design thinking and practices would suggest processes that work between two or more species perspectives or realities, by combining and synthesizing different species needs and experiences. The more literal approach to biomimicry, as one that merely mimics how natural systems and living entities survive, constitutes interspecies design, or a timid version of transpecies design. I would describe more sophisticated versions of biomimicry, as outlined by Henry Dicks in his contribution to this anthology, as one type of transpecies design. The more nuanced version of biomimicry constitutes a transpecies design method in so far as it focuses on integrating the myriad ways in which nature can serve as a model, measure, and mentor for design goals and outcomes. Realizing the central appeal Benyus makes for a biomimetic world is ultimately a design challenge, one that calls for manufacturing 'the way animals and plants do' by 'using the sun and simple compounds to produce totally biodegradable fibers, ceramics, plastics, and chemicals.'[22] For 'the more our world looks and functions like this natural world, the more likely we are to be accepted on this home that is ours, but not ours alone.'[23] Benyus's biomimetic method dismantles the dominant position humans hold in the hierarchy of moral worth, tapping into an ethical outlook on life that is reminiscent of deep ecology.

Philosopher and founder of the deep ecology movement Arne Naess defended a radical ethic that maintains all living beings carry the same moral worth.[24] If we take Naess's biocentric ethic as the launchpad for design discourses and practices, what implications would this have? With the simple claim that human lives are no more valuable than other lives on earth, Naess's ethic calls for a fundamental shift in how human beings situate themselves in the world and the world in themselves, presenting a transpecies challenge for design: human design can no longer limit species diversity and minimize the intrinsic value of other-than-human species for human gain, with the one caveat being unless the design in question is meeting the basic survival needs of humans. Transpecies design honors biodiversity as a common heritage, one we share with other species. Rather than presenting the human experience and flourishing as the goal of design, transpecies design takes the inextricable linkages connecting animals and plants as both the point of departure and the end goal of design. As such, transpecies design moves beyond human experience and behavior, serving as the fundamental ingredient for how design decisions are made and the direction design processes take. Transpecies design, as I understand it, adopts a holistic approach to species interaction, whereby the specificities of each species come into relation with one another, then in their interaction something new emerges such that their interaction changes both.

Transpecies design attends to how species interact and the relations of mutual enrichment central to a thriving living system. By integrating the needs and experiences of different species into design thinking and practices, the goal is to achieve a more holistic design approach inclusive of many species. There are two fundamental principles driving transpecies design. The first is to move design practices and thinking beyond an anthropocentric viewpoint. The second, and not unrelated to the first, is to understand the differences between species as positive. How might species differences be structured by a difference-in-itself in place of a negative difference structured dualistically. For example, if the difference between *Homo sapiens* and *Bombus impatiens* (common eastern bumblebee) is broached in terms of how the specificity of each species becomes a source of resonant flourishing, rather than in absolute dualistic terms, by focusing on what each species does not share in common with the other, then this would have significant implications for how built environments can be designed and put to work in the service of a variety of species. Attending to the differences between species as a matter of being a difference-in-itself, as opposed to a negative difference (for example, *Homo sapiens* is a human because it is not a *Bombus impatiens*), prompts designers to reflect upon and address the conditions needed for specific and shared flourishing. The goal is to create environments of reciprocal benefit that spatially and temporally enables both bumblebee and human being to thrive.[25] Here I am thinking of the 1.45-mile-long Highline Park in New York City that was created in 2009 on an abandoned railway. This highly trafficked greenbelt on the West side of New York 'contains more than 110,000 plants within 500 vascular plant species' and its flowering plants attract multiple species of bumblebees each year.[26] The Highline Park reintroduces natural elements that become the materials animals use to make other materials, such as wax, silk, and paper, which they in turn use to build shelters, capture food and store it, as well as for communication.

Transpecies design seeks to overcome what Donna Haraway has described as 'human exceptionalism and bounded individualism' by decentering the human in design.[27] In this regard, transpecies design honors biodiversity as a common heritage, one we share with other species. Instead of presenting human experience and flourishing as the goal of design, it takes the inextricable linkages connecting a multiplicity of animals and plants as both the point of departure and the end goal of design.

Moving design beyond human experience and behavior serving as the fundamental ingredient for how design decisions are made and the direction design processes take, transpecies design engages the substantive realities of a multiplicity of species and the ecosystems and biomes in which their wellbeing is imbricated, as both design content and form, harnessing the affective capacity of design to maximize deep flourishing using the following 3 Rs: Regeneration, Restoration, and Reconciliation.

Regeneration, Restoration, and Reconciliation

Reframing design as regenerative shifts our thinking beyond doing 'less bad' and onto design practices that aspire to restore and give back to the earth. It is a biome-in approach to design that entails building or designing-in biodiversity. This leans on recent work that encourages us to not only create buildings that move beyond net zero and become net positive, but to also create built environments that add ecological value and spur on biodiverse flourishing. Regenerative design is like a form of reverse engineering, recovering the natural spatial organizations and temporal processes of indigenous physical systems and elements to achieve a more holistic design outcome.[28] One example would be Michael Van Valkenburgh's design of Harold Simmonds Park in Dallas. It involves rewilding the Trinity River floodway. The design aspires to reconnect the river to the public through a series of trails, bike paths, recreational areas, and water ways designed to flood during periods of heavy rainfall. Sensitive to the different soils, sediments, hydrological features of the Trinity River basin, climatic variation, human activities, and the many different species that depend upon a healthy watershed system, Van Valkenburgh's flexible design adopts a biome-in approach by restoring the different watershed ecosystems and maximizing the ecological synergies between these.

Restoring ecosystems in priority areas yields the best outcomes for biodiversity and climate mitigation. Identifying which areas will yield the highest benefits when restored avoids having to make disadvantageous compromises between outcomes. In 2020 Bernardo Strassburg and his team at the Rio Conservation and Sustainability Science Centre in Rio, Brazil identified a series of priority areas worldwide for ecosystem restoration. The team discovered that 465 billion tons of CO_2 would be sequestered and more than 70% of projected extinctions of plants and animals in the world could be counteracted by restoring a mere 30% of identified global priority areas.[29] Using a series of evaluative criteria that prioritized the importance of spatial planning in optimizing ecosystem restoration, Strassburg and his team identified 'priority areas for restoration across all terrestrial biomes' globally, estimating the costs and benefits of each.[30] They discovered that the 'inclusion of several biomes is key to achieving multiple benefits.'[31] Furthermore, using their multicriteria approach they found that 'cost effectiveness can increase up to 13-fold when spatial allocation is optimized.'[32]

The third principle of transpecies design – reconciliation – aspires to create structures of mutual species flourishing and healing. Here I am drawing on the logic of reconciliation frameworks that have been developed in Australia and Canada in an effort to heal the past wrongs colonists waged against indigenous peoples. The *Truth and Reconciliation Report* that came out of the Canadian Truth and Reconciliation Commission states: 'Reconciliation is about establishing and maintaining a mutually respectful relationship between Aboriginal and non-Aboriginal peoples in this country. In order for that to happen, there has to be awareness of the past, an acknowledgement of the harm that has been inflicted, atonement for the causes, and action to change behaviour.'[33]

Building with an awareness of past harms and with a view to redressing those harms to spur on transformative action, I expand the logic of the Truth and Reconciliation Commissions into the spaces of design thinking and practice in an effort to recognize the enormity of suffering other-than-human-species have incurred as a result of accumulated environmental and social traumas that have taken place over the past two centuries.[34] Reconciliation used in this way approaches human settlements as platforms for healing and as structures for mutual flourishing. For example, Stefano Boeri, *Vertical Forest* (2014) in Milan. The two condo buildings that Boeri has created employ a new building typology using outward facing structures that provide a home to thousands of different species and one that enables humans and other-than-humans to live together. As Boeri describes it, 'the vertical forest is the prototype building for a new format of architectural biodiversity which focuses not only on human beings but also on the relationship between humans and other living species.'[35] The buildings are situated on 3,000 square feet of urban land, providing approximately 100,000 square feet of single family housing and the equivalent of 30,000 square meters of woodland and undergrowth. In addition to single family housing the buildings are home to:

> 800 trees (480 first and second stage trees, 300 smaller ones, 15,000 perennials and/or ground covering plants and 5,000 shrubs, providing an amount of vegetation equivalent to 30,000 square meters of woodland and undergrowth, concentrated on 3,000 square meters of urban surface … Unlike "mineral" facades in glass or stone, the plant-based shield does not reflect or magnify the sun's rays but filters them thereby creating a welcoming internal microclimate without harmful effects on the environment. At the same time, the green curtain "regulates" humidity, produces oxygen and absorbs CO_2 and microparticles…[36]

Introducing wildlife infrastructures into the concrete jungles of urbanity facilitates both humans and other-than-humans to flourish. Another example would be Terreform ONE's eight story monarch butterfly habitat, which consists of a vertical meadow and directly takes on the challenge of building back habitats for the rapidly declining monarch butterfly population. One of the many migration paths North American monarchs take on their way to hibernate through the winter months in Mexico is through New York state. The building-as-monarch sanctuary integrates biodiversity infrastructure into the skin of the building.

Any reconciliation involves repairing past harms. In the context of design, the principle of repair introduces us to the field of biodesign: designing with living materials and organic processes in a relationship of mutual respect with a view to growing buildings and maybe even eventually arriving at a place where buildings might heal themselves one day. The *mycelium chair*, by Eric Klarenbeek, combines mycelium, a filament structure that fungi uses to grow, with water and powdered straw to 3D print a chair that is in turn strengthened over time as the fungus continues growing inside of the structure. The chair literally grew into a chair as a result of the mycelium growing throughout the 3D printed straw core, which in turn transformed the water into a solid. Fungi growth is halted when the chair is dried out.[37]

Invoking the principle of repair as central to transpecies design draws on recent research into growing buildings that might one day autonomously heal themselves. The idea of using metabolic materials in the way that researchers Rachel Armstrong and Mark Bedau are doing opens up tremendous possibilities for architecture. It pushes us into the domain of creating building materials that have the living properties of a metabolism, and yet are not alive. Armstrong explains: 'Metabolic materials work with the energy flow of matter and systems using a bottom up approach to the construction of architecture.'[38] She gives the example of protocell droplets which can move, sense (responding to vibration), and modify the chemistry of their environment. 'Protocell technology,' Armstrong explains, 'could stop the city of Venice from sinking on its soft geological foundations by generating a sustainable, artificial reef under its foundations.'[39]

Holistically integrating the principles of regeneration, restoration, and reconciliation into design thinking and practices, with a view to moving beyond an anthropocentric design framework could dramatically transform the types of land cover human beings introduce on earth. Rather than dominating the earth and its biodiversity, transpecies design is ultimately a practice of commoning, embracing the idea of enrichment as a common right that all species hold.[40] In short, transpecies design could move human built environments away from a speciesist and ultimately mono-functional terrestrial surfacing to biodiverse rich environments that enable a multiplicity of species to thrive. Incorporating different classes of vegetation that function as habitats and sources of flourishing for other-than-human species, as much as they serve *Homo sapiens* flourishing, must be the future of design if human beings are going to have any hope of halting or even slowing the unprecedented rate of species now facing extinction. Ultimately, there is no human flourishing in the absence of biodiversity.

Notes

1 Intergovernmental Science-Policy Platform on Biodiversity and Ecosystem Services (IPBES), *Summary for Policymakers of the Global Assessment Report on Biodiversity and Ecosystem Services of the Intergovernmental Science-Policy Platform on Biodiversity and Ecosystem Services*, S. Díaz et al., eds. (Bonn: IPBES Secretariat, 2019).
2 Robert Hazen, *The Story of Earth: The First 4.5 Billion Years, from Stardust to Living Planet* (New York: Penguin Books, 2012); Andrew Knoll, *A Brief History of Earth: Four Billion Years in Eight Chapters* (New York: Harper Collins, 2021).
3 Yuval Noah Harari, *Sapiens: A Brief History of Humankind* (New York: Harper Collins, 2015), 13–14, 20.
4 Telmo Pievani, "The Sixth Mass Extinction: Anthropocene and the Human Impact on Biodiversity," *Rendiconti Lincei*, no. 25 (2014): 85, https://doi.org/10.1007/s12210-013-0258-9.
5 United Nations, "Introduction," in *The World at Six Billion* (United Nations: 1999), 3, https://www.un.org/development/desa/pd/sites/www.un.org.development.desa.pd/files/files/documents/2020/Jan/un_1999_6billion.pdf.
6 "Day of 8 Billion," United Nations, accessed January 4, 2023, https://www.un.org/en/dayof8billion#:~:text=On%2015%20November%202022%2C%20the,nutrition%2C%20personal%20hygiene%20and%20medicine.
7 Anders Krough, *State of the Tropical Rainforest: The Complete Overview of the Tropical Rainforest, Past and Present* (Oslo: Rainforest Foundation Norway, 2020), 3, https://d5i6is0eze552.cloudfront.net/documents/Publikasjoner/Andre-rapporter/RF_StateOfTheRainforest_2020.pdf?mtime=20210505115205.
8 Daniel O'Neill et al., "A Good Life for all Within Planetary Boundaries," *Nature Sustainability*, no. 1 (05 February 2018): 88–95, https://doi.org/10.1038/s41893-018-0021-4.
9 "Earth Overshoot Day," Global Footprint Network, accessed January 6, 2023, https://www.overshootday.org/.
10 Michael Borucke et al., "Accounting for Demand and Supply of the Biosphere's Regenerative Capacity: The National Footprint Accounts' Underlying Methodology and Framework," *Ecological Indicators*, no. 24 (2013): 519.
11 Borucke et al., 519.
12 "Data and Methodology," Global Footprint Network, accessed January 7, 2023, https://www.footprintnetwork.org/resources/data/.

13 "Data and Methodology"; Borucke et al., "The Biosphere's Regenerative Capacity," 518; "How Many Earths? How Many Countries?," Global Footprint Network, accessed January 6, 2023, https://www.overshootday.org/how-many-earths-or-countries-do-we-need/.
14 Joel Cohen, "Population Growth and Earth's Carrying Capacity," *Science*, no. 269 (1995); Vaclav Smil, *Enriching the Earth: Fritz Haber, Carl Bosch, and the Transformation of World Foods* (Cambridge: MIT Press, 2001). This situation presents another wicked problem concerning our dependence on biotechnologies, such as Genetically Modified Organisms (GMOs), to feed the current global population. The introduction of GMOs threatens natural species and biodiversity by reducing DNA diversity amongst species.
15 Ibid., The Global Footprint Network.
16 Ibid.
17 "Urban Development," World Bank, accessed January 24, 2023, https://www.worldbank.org/en/topic/urbandevelopment/overview.
18 Erle Ellis and Navin Ramankutty, "Putting People on the Map: Anthropogenic Biomes of the World," *Frontier Ecological Environments*, no. 6 (2008): 439.
19 Ellis and Ramankutty, 439.
20 Peter Singer, *Animal Liberation: A New Ethics for our Treatment of Animals* (New York: Random House, 1975), 6.
21 Singer, 6.
22 Janine Benyus, *Biomimicry: Innovation Inspired by Nature* (New York: Harper Collins, 1997), 2.
23 Benyus, 3.
24 Arne Naess, "The Shallow and the Deep, Long Range Ecology Movement: A Summary," *Inquiry: An Interdisciplinary Journal of Philosophy* 16, no. 1–4 (1973): 95–100.
25 The binomial nomenclature used to scientifically categorize different species.
26 Richard Stalter, Jingjing Tong, and James Lendemer, "The Flora on the High Line, New York City, New York: A 17-Year Comparison," *The Journal of the Torrey Botanical Society* 148, no. 3 (09 September 2021): 243–251, https://doi.org/10.3159/TORREY-D-21-00007.1.
27 Donna Haraway, "Tentacular Thinking: Anthropocene, Capitalocene, Chthulucene," in *Staying with the Trouble: Making Kin in the Chthulucene* (New York: Duke University Press, 2016), 30.
28 Biome= the vegetation, climate, soil, water, and wildlife making up a habitat or large area. There are 5 kinds of biomes – forest, grassland, aquatic, desert, and tundra. Image: Michael Van Valkenburg, Dallas rewilding project using the freshwater biome of the river system.
29 Bernardo B. N. Strassburg et al., "Global Priority Areas for Ecosystem Restoration," *Nature*, no. 586 (14 October 2020): 724–729, https://doi.org/10.1038/s41586-020-2784-9.
30 Strassburg et al., 724–729.
31 Strassburg et al., 724–729.
32 Strassburg et al., 724–729.
33 Truth and Reconciliation Commission of Canada, *Honouring the Truth, Reconciling for the Future: Summary of the Final Report of the Truth and Reconciliation Commission of Canada* (Ottawa: Truth and Reconciliation Commission of Canada, 2015), 6–7, https://publications.gc.ca/site/eng/9.800288/publication.html.
34 Complexity – empowering natural systems Healing – making amends Tartu Nature House by KARISMA architects (2013) Estonia - "a symbiosis between a zoo, botanical garden, and school inspired by a tree stump" (KARISMA)
35 "Vertical Forest," Boeri: Stefano Boeri Architects, accessed March 2, 2023, https://www.stefanoboeriarchitetti.net/en/project/vertical-forest/.
36 "Vertical Forest."
37 Marcus Fairs, "Mycelium Chair by Eric Klarenbeek is 3D-Printed with Living Fungus," *deezen*, October 20, 2013, https://www.dezeen.com/2013/10/20/mycelium-chair-by-eric-klarenbeek-is-3d-printed-with-living-fungus/.
38 Rachel Armstrong, "Self-Repairing Architecture," *NextNature* (blog), June 24, 2010, https://nextnature.net/story/2010/self%E2%80%93repairing-architecture.
39 Armstrong; Mark A. Bedau, "Living Technology Today and Tomorrow," *Technoetic Arts* 7, no. 2 (2009): 199–206.
40 Adrian Parr Zaretsky, "Commonism," in *Birth of a New Earth: The Radical Politics of Environmentalism* (New York: Columbia University Press, 2018), 91–119.

3
TRANSPECIES DESIGN AND BIOMIMICRY

Henry Dicks

The word "transpecies" has been employed in a number of contexts recently and it was perhaps only a matter of time before it became linked to design. Nevertheless, exactly what "transpecies design" might be remains unclear, for it would not yet appear to have been the object of any sustained attempt at conceptualization. Drawing on transpecies approaches in other fields, especially urban theory and psychology, the first part of this chapter will argue that these approaches share common methodological, ontological, and ethical assumptions, all of which may potentially be transferred into the field of design. This will then pave the way for the second and third parts, in which I argue that articulating the notion of transpecies design with biomimicry may make it possible to respond to some of the recent criticisms that have been levelled at biomimicry, while also showing the limitations of transpecies design, understood and practiced as something distinct from biomimicry.

Conceptualizing Transpecies Design

The expression "transpecies design" is not yet well established in the academic literature. It is nevertheless possible to conceptualize it by drawing on transpecies approaches in other contexts, notably urban theory and psychology. In what follows, I argue that all transpecies approaches have three important aspects in common, relating to: methodology, ontology, and ethics.

The key methodological claim made in transpecies urban theory is that the city should no longer be considered simply as a site for human habitation and activity, for other species are also active inhabitants of the city, and it is thus possible to study the city as a site of transpecies interactions between humans and other species.[1] What, following Andrew Pickering, one might call the "unit of analysis" of transpecies urban theory is neither human beings, nor other species, but interactions between the two.[2] In an important sense, then, transpecies urban theory is also transdisciplinary, and not simply interdisciplinary, for it is not simply a question of bringing together separate disciplinary studies of humans (anthropology, sociology, etc.) and of other species (zoology, ecology, etc.), but rather of studying the interactions themselves, that is, in the context of urban theory, the ways in which the intertwining agencies of humans and other species participate in making the city what it is. Something similar applies also to transpecies approaches in other disciplinary contexts. The unit of analysis of transpecies psychology, for example, is the interactions that occur between the psyches of humans and other species.

The second important aspect of transpecies approaches is ontological. Drawing on postmodernist and posthumanist criticisms of human-nature dualism, a first key claim made is that other species possess important ontological characteristics that traditional dualisms reserve for humans. The most prominent of these is agency. If transpecies urban theory studies the interactions that occur between humans and other species in the city, this can only be because other species are agents whose actions intertwine with those of human agents. Similarly, a key claim of transpecies psychology is that many other species possess a psyche and, because of that, may partake in psychological interactions with humans, whether negative, as occurs when human-induced trauma leads to the development of psycho-pathologies in other species, or positive, as occurs when humans play a therapeutic role, helping trauma inflicted on the psyches of other species to heal.[3] Further, the word "transpecies" also works to undermine the ontological gap between humans and other species in a converse way: contrary to traditional humanisms, it implies a biological conception of humans as a "species," albeit one which may retain certain distinctive characteristics.

The third important component of transpecies approaches is ethical. Perhaps unsurprisingly, given the terminology involved, that to which transpecies approaches are often most explicitly opposed is "speciesism"[4] – a concept forged by Peter Singer in the context of animal ethics to denote discrimination against other beings solely on the grounds that they belong to a certain species.[5] But transpecies approaches amount to more than simply anti-speciesism, for "transpeciesism" seeks also to provide some positive ethical content. So what is that? There can be little doubt that transpecies approaches are opposed to anthropocentrism in ethics: the view that only humans have moral standing. And yet it seems fair to say that they have not embraced any of the main non-anthropocentric positions put forward in environmental ethics: zoocentrism (or sentientism), which holds that only (sentient) animals have moral standing; biocentrism, which holds that all living beings have moral standing; and ecocentrism, which holds that ecosystems have moral standing. The reason for this, I suggest, concerns the standard opposition between intrinsic and instrumental value current in mainstream environmental ethics. As a general rule, these two value concepts have been mobilized as follows: either we are called upon to respect the intrinsic value of nature and so to leave it alone and not instrumentalize it; or it is assumed that nature does not possess intrinsic value, though it may possess instrumental value, in which case we may exploit it however we choose. From this perspective, the only interactions we may have with other species – or at least the only material ones – involve instrumentalizing them.

Where transpecies ethics breaks with these stark alternatives is in assuming that it is possible to interact with other species in such a way that there occurs what one might call "mutual flourishing." We may, in other words, benefit from our interactions with other species, just as they may benefit from their interactions with us. Other species, from this perspective, could be treated as a means, but they could never – to paraphrase the second formulation of Kant's categorical imperative – be treated "as means only."[6] Conversely, while humans could potentially serve as a means for other species, they would at the same time have to be respected as ends – a view which means humans cannot be sacrificed for the benefit of other species, as may occur in the case of eco-fascism. Further, mirroring the methodological shift described above, the ethical focus of transpecies ethics would be the interactions themselves. So, whereas modern dualism cuts humans off from other species, transpecies approaches not only study interactions between species but further maintains that what is ethically important is above all to foster and cultivate forms of interactions with other species that foster mutual flourishing.

*

Drawing on the above analyses, let us now turn our attention to transpecies design, starting with its methodology. Traditionally, it has been assumed that humans alone are designers, with other species being at most parts of human designs. Modern agricultural systems, for example, feature other species – corn, soy, etc. – as part of the design, but these other species do not actively participate in the design process. In the case of transpecies design, by contrast, other species would be active participants. But how exactly might that work? Before answering this question, let us note that another important methodological feature of traditional conceptions of design is that there is some sort of planning phase during which the designer plans out what it is they wish to realize before actually realizing it, and which traditionally involves producing sketches, blueprints, models, prototypes, and the like. Now, one pitfall I think must be avoided when conceptualizing transpecies design is seeing other species as potential co-designers, in the sense of being able to come up with plans, prior to the execution phase, such that we might be able to combine these plans with our own. Karl Marx famously remarked that "[w]hat distinguishes the worst architect from the best of bees is this, that the architect raises his structure in imagination before he erects it in reality."[7] But while Marx may perhaps be right that this ability to imagine a design before realizing it is unique to humans, there are two important qualifications that must be made to his claim in the context of transpecies design.

The first is that the design process may involve transpecies interactions, by which I mean that, rather than being fully planned out by humans before being erected in reality, the entire design process may involve multiple twists and turns depending on the interactions that occur with other species. In place, then, of a simple two-stage process, involving imagining and realizing, planning and execution, there would be a multi-stage process in which the design first conceived by humans undergoes multiple modifications depending on the interactions that occur with other species. The human designer, from this perspective, would need to be responsive to the actions of other species, and so cease to think of themselves as the sole agent in the design process.

This brings me to the second qualification. When Marx says that the architect "raises his structure in imagination before he erects it in reality," it is seemingly implied that the architect imagines the structure *in full*, as it were, such that erecting it in reality becomes a relatively trivial process of transposing a fully conceived design into material reality. Further, were one to adopt a Platonic logic of design, one could even say that, far from realizing or completing what was previously only imagined, the material product introduces imperfections to a design that was perfect only in the imagination. But whether the design logic is that of Marx or Plato, the common ground is the idea that what the final product will be is *determined in advance*, and the human designer thus retains full control over the end-product. In the case of transpecies design, by contrast, what the resulting design will be exactly is *open-ended*, for it depends on the transpecies interactions. The human designer will of course need a rough idea of what they are trying to achieve, but not only will the precise details of their design remain open, but even the rough idea they began with could potentially change, depending on the interactions that take place with other species.

A simple example may help understand what I mean. Imagine someone has acquired a plot of land which they wish to turn into an organic farm. Now, they may have a rough idea of which plants they think will grow well, how these plants might interact with one another and with other wild species, what techniques they might use to deter pests, how they might improve soil fertility, and so on, and they will thus begin planting and farming accordingly. But not only may some of the non-human species involved in the initial design likely act somewhat differently from expected, but the designer may also have to contend with various wild species of which they were previously little aware. In such circumstances, rather than trying to force through their initial design, the designer would modify the design, for example by planting different crops, modifying the soil, or seeking to attract a species of wild predator they had not initially realized was present in the vicinity. What the farm looks like at any given moment in time will thus be a result of interactions between the human designer and other species, and its design would thus be an open-ended process in which other species actively participate. This approach may be contrasted with that of industrial farming, in which a single model, involving purchased seeds, fertilizers, pesticides, and so on, is applied, with any difficulties presented by other species (e.g., pesticide resistance) being met simply by reinforcing the basic model (e.g., stronger or alternative pesticides).[8]

Turning now to the question of ontology, transpecies design clearly sees other species as agents. It goes beyond this, however, inasmuch as it introduces non-human agency into a new realm: that of design. Non-humans, from this perspective, are not just active participants in reality, but active participants in the design of new realities. Conversely, the agency of human designers would also be reconceptualized. No longer could human designers be qualified simply as active, for they would also be reactive and interactive, and no longer could human design involve perfection in the imagination before erection in reality, for it would instead involve a series of interactions between a far from perfect imagination and a complex and ever-changing reality. But while there is an important sense in which transpecies design reduces the ontological divide between humans and other species, this is not to say the divide can be eliminated completely. Other species would participate actively in the design process, but without being co-designers, by which I mean that they would not be actively involved in the imaginative work of *planning* future realities. And while humans may relinquish full control of the design process, they would not relinquish their unique status as designers.

Turning now to the ethical dimension of transpecies design, it is clear that it would also adopt the ethic of mutual flourishing common to other transpecies approaches; any transpecies design would seek to enable both its human and its non-human participants to flourish. Other species, it follows, could not be instrumentalized, in the sense of being reduced merely to a means, for they would also need to be respected as entities possessing a good of their own. And yet transpecies design would aim for more than just respecting the intrinsic value of other species, for it would also seek out opportunities for positive interactions involving mutual flourishing between human and non-human species. In the case of the organic farm presented above, creating habitat for wild species of benefit to the farm – predators, pollinators, etc. – would be an obvious example.

Transpecies Design and Biomimicry

An obvious limitation of transpecies design is that, taken on its own, it does not have anything much to say about the sorts of designs we should be trying to achieve. We know transpecies designs should involve other species as active participants and that they should aim to achieve mutual flourishing, but we remain in the dark about the contribution of the human designer, that is to say, the process of making plans, blueprints, models, and so on, that presumably remains a necessary part of transpecies design.

One way in which this limitation might be addressed is through biomimicry, an increasingly popular design strategy that involves imitating, emulating, and learning from nature.[9] The initial design of an organic farm, for example, could be based – even if only quite loosely – on the native ecosystem. This raises the intriguing possibility that other species could be involved in the planning phase after all, namely be providing the initial plans, templates, or models that the human designer would then seek to transfer over into material reality. Moreover, in keeping with the idea that biomimicry involves learning from nature, it could perhaps be thought that other species may take on the role of a very specific type of agent: a mentor.[10] In the case of, say, the Shinkansen 500 series, a high-speed Japanese train whose form is based on the beak of the kingfisher, it is the kingfisher that holds this role. It is questionable, however, whether the kingfisher is playing an *active* role here. Indeed, the agency of the kingfisher seems irrelevant to the process of taking the form of its beak as a model for a biomimetic design. We may be taking the kingfisher as a mentor, but this is not a role that the kingfisher itself is acting out.

There are cases of biomimetic design, however, that allow for nature to participate not only as mentor during the planning phase, but also in the material realization of the design itself. I am thinking especially of those cases in which the human designer is seeking to imitate, emulate, and learn from natural ecosystems, as occurs in such fields as ecological engineering,[11] permaculture,[12] or analog forestry.[13] In these instances, other species do not simply inspire a design from which they themselves are absent; they participate also in the design itself. But are they *active* participants? If the human designer aims to design the system *in full* before erecting it in reality, then the various non-human species involved in the design are being treated simply as passive materials, and not as beings with whose agency the human designer seeks to interact. The danger remains, then, that nature may only participate in biomimetic designs in ways that preclude it taking on an agential role in the design process.

The state of affairs described above may go some way to explaining some of the criticisms that have been levelled at biomimicry. Karen Barad has argued that biomimicry – at least as theorized by Benyus – presumes that "there is a pure nature separate from culture."[14] Other species may provide the initial blueprints for biomimetic designs, but they are not active participants in the design process. In contrast, then, to Barad's own "agential realism," which seeks to study the entangled agencies of humans and non-humans[15] – and thus shares much in common with the transspecies approaches I conceptualized above – the traditional assumption that humans are the only *active* participants in design would appear to be upheld.

Michael Fisch makes a similar criticism. Taking objection to what he describes as "Benyus's insistence on the establishment of clear and impervious borders between nature and technology,"[16] Fisch argues that the work of Neri Oxman – as exemplified by her work "Silk Pavilion" – constitutes a preferable alternative. Oxman's original intention, Fisch tells us, was a classic instance of traditional biomimetic design: to take the silkworm cocoon as a model for a concept piece of human architecture to be realized using a 3D printer.[17] But, as this approach did not work, Oxman instead sought to involve the silkworms themselves in the design, using a swarm of them as "a kind of bio-3D-printer," such that the human designer and the silkworm ultimately partake in a form of "relational becoming,"[18] understood as an open-ended process in which what is ultimately realized is not fully determined in advance but instead emerges through the interactions of the human designer and the silkworms – a clear instance of what I described above as transspecies design. Fisch also draws attention to Oxman's concern to treat the silkworms ethically, which he frames in terms that are strongly reminiscent of the ethics of transspecies design I presented above: "the central question in determining the ethical integrity of the project becomes the degree to which the project enables the flourishing and further becoming of its various participants."[19] In short, it would seem that Oxman may have made a transition from biomimetic to transpecies design. But are we to conclude, then, that biomimicry and transpecies design are opposed, perhaps even that we must choose between one and the other? Or is there perhaps scope for articulating the two, and in such a way that the limitations of each approach, taken individually, might be overcome?

Biomimetic Design and the Question of Temporality

With a view to answering these questions, let us note that the above criticisms of biomimicry presuppose its adherence to the traditional temporality of design, that of design and execution, planning and realization. Oxman's initial plan for Silk Pavilion, for example, involved taking the silkworm cocoon as a model for an architectural design that was supposed to be simply "printed out" in 3D. But is there not scope for introducing the open-ended multi-stage temporality of transpecies design into biomimicry? And, if so, how might that work? With a view to answering these questions, let us take a closer look at biomimicry.

I noted earlier that nature participates in biomimetic designs through providing models for human designers. But biomimicry does not call on us only to take nature as model, but also as measure. Beyond simply imitating or taking inspiration from natural systems, this second principle calls on us also to measure our designs against nature's "ecological standard."[20] Now, the way I suggest that biomimicry may be articulated with the open-ended multi-stage temporality characteristic of transpecies design is through the *iterative interplay of the principles of nature as model and nature as measure.*

Let's return again to the example of an organic farm. In this instance, the initial biomimetic design would likely be based on models abstracted from the native ecosystem and may include such things as the use of local crop varieties, polyculture, natural predators, and nutrient cycling. Once this initial design is realized, it would then be possible to apply the principle of nature as measure, that is to say, it would be possible to measure the performance of this system against that of the native ecosystem. If, for example, there is significant soil erosion, if extensive irrigation is still required, if natural predators fail to deal with pests, and so on, then it may be necessary either to return to the models already abstracted from the native ecosystem and develop a different system on their basis, or to abstract somewhat different models from the native ecosystem. A new system could then be designed and realized, with its performance once again being measured against that of the native ecosystem. In place, then, of a simple two-stage temporality of design and execution, there would be a complex multi-stage temporality involving the iterative interplay of the principles of nature as model and nature as measure.

This multi-stage temporality lends itself to articulation with that of transpecies design. After all, if the design were to deviate significantly from nature's ecological standard, this would typically be because the species involved in the design were not acting as expected. Imagine, for example, that the natural predators that were supposed to be eating the pests were staying away, and that pesticides were thus required to save the harvest, but at the cost of failing to meet nature's ecological standard. In this instance, further observation of the native ecosystem could allow the designer to notice that natural predators do not only require food to eat, but also a specific type of natural habitat, and this could lead the designer to plant hedgerows of native tree species near the cultivated crops. The design would thus be a result of multiple species – humans, crops, pests, predators, native tree species – in interaction, rather than being a pre-conceived human plan that involves other species only for its material realization.

Articulating these two open-ended multi-stage design temporalities provides further insight into the limitations of transpecies design, practiced as a stand-alone approach. Without the models and measures provided by nature, the only principle capable of guiding the transpecies designer is that of "mutual flourishing." But this is far too vague to help the designer develop any concrete designs. In Oxman's case, for example, a transpecies ethic clearly led her to look for ways of obtaining usable silk structures without killing the silkworms (as occurs in industrial silk production when the larvae are boiled up at the same time as the cocoons),[21] but it provides little guidance as to how any actual design process is to proceed. Moreover, were the designer to start with a natural model, such as the form of the silkworm cocoon, and then leave biomimicry behind as interactions between the designer and other species take over, the risk is that the project would fail to live up to nature's ecological standard. Introducing the principle of nature as measure, by contrast, would allow the designer of, say, a silk pavilion to compare silk production in nature with that going on in the case of their transpecies design project and then to seek to measure the ecological performance of the latter against that of the former. This would likely involve considering such issues as the provenance of the mulberry leaves on which the silkworms feed, the carbon footprint of the project, issues relating to the recycling of any silk pavilions obtained, and so on.

Articulating transpecies design with biomimicry in the manner described above also allows us to identify another potential problem with transpecies design: it risks going too far in its attempt to undermine the ontological distinction between humans and other species. Since transpecies design, at least as I conceptualized it above, has little to say about the work of imagining, planning, modelling, and the like, carried out by the human designer, the fact that other species do not contribute in this manner to the design process is easily overlooked and they may thus quite plausibly be described – as Oxman does – as "co-designers."[22] Once transpecies design is articulated with biomimicry, by contrast, the human designer takes on a very distinctive role. Other species may interact with one another, just as we may interact with other species, but it is we alone who must come up with models – through abstraction from nature in this instance – to be transferred over into material reality. So, while it is true that the dividing line between humans and other species may need to be reconceptualized, it cannot be eliminated altogether.

Conclusion

In a wonderful article of great interest to transpecies designers, Andrew Pickering contrasts two different painting styles.[23] The first is that of Piet Mondrian's famous grid-like compositions in black, blue, red, yellow, and white. The second is that of Willem de Kooning. The difference between these two styles, Pickering tells us, is that whereas Mondrian has an exact idea of what he is aiming to paint, de Kooning begins by smearing paint on the canvas, then moves it around looking for "beautiful and complex configurations,"[24] then smears on some more paint, and so on, until he is satisfied with the result. Pickering then makes the following suggestion:

> We could, collectively, try being de Kooning rather than Mondrian. We could look for the beauty, very broadly understood, natural and social, in the outcomes of our interactions with the environment, and we could try to work on and amplify that when we find it.[25]

If we substitute the ethical goal of "mutual flourishing" for the aesthetic one of "beauty," then being de Kooning and being a transpecies designer would amount to much the same thing.

But Pickering's critique of Mondrian overlooks an aspect of Mondrian's art that is at least as important as its simple two-stage temporality of design and execution. In 1905, the German philosopher of technology, Max Eyth, challenged the longstanding view of technology as imitation of nature, defending instead its "spiritual autonomy."[26] Fourteen years later, Mondrian rejected the equally longstanding view of art as imitation of nature, arguing that art results rather from the autonomous life of the human mind.[27] In keeping with this, he not only sought to turn away from natural things and paint in as abstract a way as possible, but also to eschew natural forms and colours, especially curved lines and the colour green. So, while Pickering may be right to say that "we currently treat the environment like Mondrian treated paint,"[28] this is not only because Mondrian's painting is underpinned by the maxim "resist any contingencies that may threaten your preconceived goal,"[29] but also because the goal he shared with modern industrial technology – to realize the full autonomy of the human mind by deliberately turning away from and going against nature – has made environmental destruction all but inevitable. By the same token, however, it would also be insufficient to interact with other species in the way that de Kooning – as much an abstract expressionist as Mondrian – interacted with paint. Design must be more than simply a "dance of human and nonhuman agency,"[30] for this dance must be given structure and direction by means of biomimicry, and more specifically, by means of the iterative interplay of two principles: nature as model and nature as measure.

Notes

1 Jennifer R. Wolch, Kathleen West, and Thomas E. Gaines, "Transspecies Urban Theory," *Environment and Planning D: Society and Space* 13, no. 6 (1995): 735–60; Alice Hovorka, "Transspecies Urban Theory: Chickens in an African City," *Cultural Geographies* 15, no. 1 (2008): 95–117.
2 Andrew Pickering, "Asian Eels and Global Warming: A Posthumanist Perspective on Society and the Environment," *Ethics & Environment* 10, no. 2 (2005): 31.
3 Mary Watkins and G. A. Bradshaw, "Trans-Species Psychology: Theory and Praxis," *Spring: A Journal of Archetype and Culture*, no. 76 (2007): 69–94.
4 See, for example, Watkins and Bradshaw, "Trans-Species Psychology."
5 Peter Singer, *Animal Liberation: A New Ethics for our Treatment of Animals* (New York: Random House, 1975).
6 Immanuel Kant, *Fundamental Principles of the Metaphysic of Morals*, trans. T. K. Abbott (New York: Prometheus, 1988), 58.
7 Karl Marx, "Capital: A Critique of Political Economy, Volume 1," Marxists Internet Archive, accessed October 18, 2022, https://www.marxists.org/archive/marx/works/1867-c1/ch07.html.
8 On this issue, see Pickering, "Asian Eels and Global Warming," 41.
9 Janine Benyus, *Biomimicry: Innovation Inspired by Nature* (New York: Harper Perennial, 1997).
10 Benyus, epigraph.
11 W. J. Mitsch, "What is Ecological Engineering?" *Ecological Engineering* 45 (2012): 5–12.
12 Charles Hervé-Gruyer and Perrine Hervé-Gruyer, *Permaculture: guérir la terre, nourrir les hommes* (Paris: Actes du Sud, 2014).
13 Ranil Senanayake and John Jack, *Analogue Forestry: An Introduction* (Clayton: Department of Geography and Environmental Science at Monash University, 1998).
14 Karen Barad, *Meeting the Universe Halfway: Quantum Physics and the Entanglement of Matter and Meaning* (Durham: Duke University Press, 2007), 368.
15 Barad, 26.
16 Michael Fisch, "The Nature of Biomimicry: Toward a Novel Technological Culture," *Science, Technology & Human Values* 42, no. 5 (2017): 816.
17 Fisch, 811.

18 Fisch, 812.
19 Fisch, 814.
20 Benyus, *Biomimicry*, epigraph.
21 Fisch, "The Nature of Biomimicry," 815–16.
22 Of course, if by "co-design" one means simply the involvement of stakeholders other than the human designer(s) in the design process then, as stakeholders, the silkworms are indeed co-designers. But they are not co-designers in the same sense that, say, a co-author is a co-author, for they themselves do not do any actual designing.
23 Pickering, "Asian Eels and Global Warming," 41.
24 Pickering, 41.
25 Pickering, 41.
26 Max Eyth, *Lebendige Kräfte: Sieben Vorträge aus dem Gebiete der Technik*, 4th ed. (Berlin: J. Springer, [1905] 1924).
27 Piet Mondrian, *Natural Reality and Abstract Reality: An Essay in Trialogue Form in 1919–1920* (New York: George Brazilier, 1995).
28 Pickering, "Asian Eels and Global Warming," 41.
29 Pickering, 41.
30 Pickering, 34.

4
DESIGN AGAINST EXTINCTION
Multispecies Methods and Engineered Living Materials

Mitchell Joachim

All life on earth is finely intertwined. This delicate equilibrium has been established over billions of years. As one species befalls extinction, many other species are disturbed, pushing several ecosystems in danger of collapsing. Past mass extinctions were triggered by extreme temperature changes, increasing or sinking sea levels, and catastrophic, singular events like a vast volcano discharging ash and gases or a colossal asteroid hammering the planet. We know about them because we can perceive how life has been altered in fossil evidence. What can we do? A sizeable part of Terreform ONE's laboratory/studio work includes investigating anti-extinction solutions through full-scale designs, technologies, and controlled experiments.

The present-day condition of the earth gives rise to a range of new perspectives in the environmental design and architecture spheres that take multispecies perspectives into account. These voices are frequently framed as more-than-human design research, as they critique the severely anthropocentric practices that spawned environmental degradation and recognize the complex nature of humans and nonhumans. Terreform ONE reflects on this agenda. We utilize a design studio/laboratory model as a salient endeavor to translate often convoluted and abstract dialogues from multispecies directives into usable forms. Our aim is to construct tangible methods, plans and design tactics for architecture, urban design, and landscape that incorporate increased biodiversity. The design studio/living laboratory at the Terreform ONE, New Lab explores a range of scales and case studies within a larger research interest area of socio-ecological theory. We work to understand the role of biodiversity in design and to reestablish human-nature interactions. A vital bond is needed within architecture to link constant exchanges in multispecies design to everyday practice. What materials and biological procedures do that the best?

Can designers generate a range of living materials that have the characteristics of biological systems: self-replication, self-regulation, self-healing, ecological responsiveness, and self-sustainability? Engineered Living Materials (ELMs) are described as engineered materials composed of living cells that form or accumulate the material itself or regulate the functional performance of the material in a particular mode. The main principle is to expand the borderlines and limits of synthetic biology, materials production, biomaterials, and artificial intelligence (AI) and push their evolution into new territories.

DOI: 10.4324/9781003403494-6

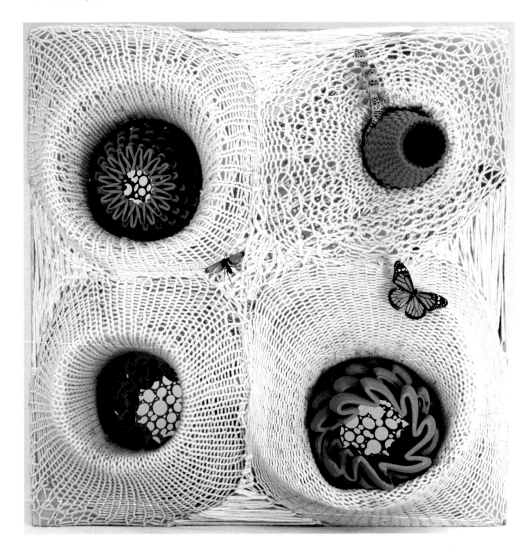

FIGURE 4.1 Multispecies wall panel prototype with woven jute and 3d printed clay habitats.

Biomaterials have evolved from inert resources that lack interaction with the body to biologically active, instructive materials that host and provide signals to adjoining cells and tissues. ELMs contain living cells for responsive function and polymeric matrices for scaffolding function and, thus, can be designed as active and reaction biomaterials. In this chapter, we discuss ELMs that incorporate microorganisms as the living, bioactive component. Microorganisms can provide complex responses to environmental stimuli, and they can be genetically engineered to allow user control over responses and integration of numerous inputs. The engineered microorganisms can either generate their own matrix, such as in biofilms, or they can be incorporated in matrices using various technologies, such as coating, 3D printing, spinning, and microencapsulation (Figure 4.1). We highlight biomedical applications of such ELMs, including biosensing, wound healing, stem-cell-based tissue engineering, and drug delivery, and provide an outlook to the challenges and future applications of ELMs.

Terreform ONE is exploring the artistic, structural, and remediated applications of ELMs in plant systems at the edges of computational design. Our projects marry advanced digital fabrication with indigenous woody plant grafting methods to push the boundaries of regenerative building material and structural design at the scale of municipal architecture. The key focus of our projects involves ELMs in which grown native trees and a green wall matrix will be guided by a technologically sophisticated parametric reusable scaffolding. Over time, the two reciprocally supportive systems intertwine and merge into a single plant-technical composite structure. Habitat inserts made from 3D-printed earthen clay and other biomaterials are embedded into the architecture, creating a vertical ecosystem that houses a variety of flora and fauna and demonstrates the possibility of cohabitation with other species.

Intrinsic to our practice is the conviction that biology is technology. Nature has evolved an ability over epochs to allocate assets, conserve energy, and regenerate itself. Our projects aim to channel nature's intelligence to develop new, restorative materials and building processes that perform ecosystem services by actively filtering the air, water, and soil and creating opportunities for life to thrive in the built environment. Our structures are closely monitored with miniaturized electronics, time-lapse video, soundscape ecology microphones, and environmental sensors to illustrate the varied and rich interactions with surrounding biota. The results are publicly accessible fecund organic structures, in which the data collected from the sensors will be visualized, and communicate the intricate web of life that exists often beyond the surface of our perceptions. The intent of the work is to illustrate the artistic and ecological possibilities of working with living material to co-create a more connected, habitable, and social environment.

Despite the ecological, psychological, and economic benefits of designing with living material, there is little industry research dedicated to scaling it to a point of widespread adoption. The work at Terreform ONE developed and cataloged feasible ELM techniques and disseminated them with the public, with underserved communities in Brooklyn, and with our institutional and organizational partners. We hope that by sharing the plethora of projects and bringing in collaborative partners, we can showcase the socio-biological potentials of designing with living materials.

Objectives of Engineered Living Materials

1 To deepen our understanding of ELMs and their potential use in the built environment, specifically focusing on low embodied carbon, earthen clays, biomaterials, landscape designs, and living woody plant structures.
2 To advance the implementation of these materials through cutting-edge computational design technologies and new-fangled AI-based formats.
3 To install living structures based on the expansive research, with an array of auditory and visual sensors to monitor changes in the environment over time, and to provide real-time experiential data.
4 To share the knowledge gained from this architecture through visualizations and the development of a curriculum with a wider audience through community outreach, seminars, and institutional forums.

Investigative Activities

To achieve these objectives, we engage in a variety of research activities, including:

1. Literature review: We conduct a thorough review of existing scientific literature on ELMs and their potential use in the built environment.
2. Computational design: We use programmable design technologies and flexible software to simulate, rationalize, and optimize the use of ELMs.
3. Experimental construction: We have conducted a series of experimental projects culminating in the installations of various living elements.
4. Public outreach: We share and disseminate the knowledge gained from this research through a series of public lectures and workshops.

Collaboration and Outreach

These research projects carried out by Terreform ONE and disseminated by us become our verifiable interdisciplinary educational platform and think-tank where we share our knowledge with the broader public (Figure 4.2). We collaborate with foundations, students, and designers to advance the study of new materials and techniques for inventive/architectural expression and ecological restoration. We distribute the outcomes of our research within the art and architecture industry through lectures and workshops, pre-doctoral fellowships, and through engagement with the wider community. The message is reverse the course of current poor development practices and stop extinction.

The sixth extinction is real and is underway. We need to use all available resources to mitigate its climatic effects. This architectural design research aims to investigate the structural and remediated applications of ELMs in the built environment and their impact on biodiversity, focusing on the intersection between traditional earth-based building methods and digital innovations in fabrication. By deepening our understanding of ELMs and how they can be used to create healthier urban ecosystems, we can open up new possibilities in the field of design. Through our active public outreach platform, we share the knowledge gained from this research with a wider audience to offer new design approaches for the future. We hope the dissemination of this research becomes a partner in the development of restorative building methods that enrich flourishing multi-species ecosystems.

Project Credits: Terreform ONE, Mitchell Joachim (Cofounder/ Principal Investigator), Vivian Kuan (Executive Director), Nicholas Lynch, Avantika Velho, Claudia D'Auria, Mamoun Friedrich-Grosvenor, Grace Morenko, River Prud'Homme, Brook Boughton.

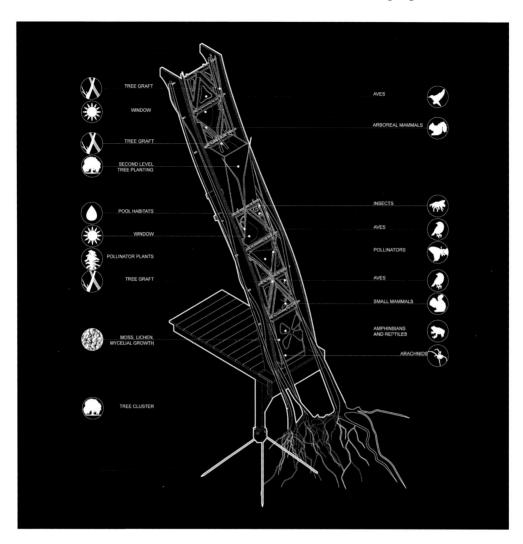

FIGURE 4.2 Diagram and photo of the Fab Tree Hab wall section that provides multispecies habitats to increase local biodiversity.

FIGURE 4.2 (*Continued*)

5
GARDEN CITY TO CITY IN NATURE

A Case for the Cohabitation of Tidal Ecologies along Singapore's Urban Waterfront

Gabriel Tenaya Kaprielian

Singapore is known as a "Garden City," a title that the city-state has embodied through an extensive network of urban trees, lush city parks, and more recently green facades and roofs. While it has earned a reputation as a world-wide leader in biophilic design, these green spaces have focused primarily on human benefits including providing shade, urban heat island reduction, and capital-driven marketing. The garden landscapes rely on human control of the

FIGURE 5.1 Artificial beach on Singapore's Sentosa Island

DOI: 10.4324/9781003403494-7

FIGURE 5.2 Sign along a hiking trail on Sentosa Island.
Photograph: The author

natural environment often replacing the historical ecology that has been shrouded by layers of development. This is especially true of Singapore's shoreline, where the country has used land reclamation to expand its area by 25 percent (Figure 5.1). While the construction of new land was seen as a vital part of national building, it erased nearly all the original shoreline along with the mangrove forests, natural beaches, and coral reefs. Age-old fishing villages that coexisted with the sea and were dependent on the tidal ecosystem have been replaced with high-rise buildings, urban infrastructure, and parks. As a result, there has been an enormous loss of coastal habitat and cultural connection to the sea.

In 2021, Singapore introduced its *Green Plan 2030* to create a future urban planning vision.[1] Included in this plan is a "City in Nature" framework developed by Singapore's National Parks (NParks) that offers a rebranding from the previous "Garden City." The City in Nature framing presents an opportunity to introduce "wild" natural systems into the built environment, rather than the typical "sanitized" constructed green landscapes (Figure 5.2). Along the urban waterfront, this could mean new development that integrates coastal habitat with a hybrid soft and engineered shoreline. This would represent a dramatic shift from past coastal development built on reclaimed land with a hard engineered shoreline. Furthermore, it would require alignment with Singapore's Urban Redevelopment Authority (URA) and the Public Utilities Board (PUB), the government entities responsible for land use planning and protecting the country from sea-level rise, respectively. *A City in Nature framework poses the question of what might new typologies of architecture and urban design look like that embrace watery landscapes, increase intertidal biodiversity, and serve as a protective barrier against rising sea levels?*

In a 2019 National Day Rally speech on Singapore's bicentennial year, Prime Minister Lee Hsien Loong discussed the challenges posed by climate change during this century, describing sea-level rise as "an existential threat" to the city-state.[2] As a low-lying island with roughly a third of the land residing below 5 meters, Singapore is particularly vulnerable to sea-level rise and compound flooding. The cost of protecting the country with climate change defense is expected to top an estimated 100 billion Singapore dollars over several generations.[3] Singapore is currently planning for a sea-level rise of 1 meter by 2100, while acknowledging that waters could rise to 4 or 5 meters with compound flooding from high tide, storm surge, and land subsidence.[4] At the same time there are ambitious plans for future waterfront development. This includes moving one of the most active ports in the world to the west coast of the island, redeveloping the Greater Southern Waterfront, creating a "long island" on the east coast, and expanding the island of Pulau Tekong by prototyping a polder system. These projects will once again redefine Singapore's shore and offer the potential to do it in a way that protects against coastal flooding by integrating new resilient development with tidal habitats. This could take the approach of nature-based solutions that are described as "sustainable planning, design, and environmental management and engineering practices that weave nature features or processes into the built environment to promote adaptation and resilience" to "combat climate change" and "reduce flood risk."[5]

While protective barriers, polders, and land reclamation are conventional approaches to shoreline development used in Singapore, these may not be long-term solutions as they contribute to climate change and loss of biodiversity. A recent report by the Intergovernmental Panel on Climate Change (IPCC) shows climate breakdown is happening more rapidly than expected and that the window to act is closing fast. The report highlights "the interdependence of climate, ecosystem and biodiversity, and human societies and integrates knowledge more strongly across the natural, ecological, social and economic sciences than earlier assessments."[6] By coupling biodiversity loss with threats of climate change such as sea-level rise, the report places prominence on solutions that connect the built and natural environment. The report states that "risks from sea-level rise for coastal ecosystems and people are very likely to increase tenfold well before 2100 without adaptation and mitigation action."[7] This is a pressing issue for Singapore, as an equatorial country it may experience rising sea levels up to 20 percent above the global average.[8]

Many of the challenges facing Singapore along its urban waterfront are shared by coastal cities around the world. As sea levels rise and compound flooding from more severe storm events threaten coastal development, this also presents an opportunity to reconsider the relationship between land and sea. The traditional perspective of a static and fixed shoreline boundary separating city and water creates an artificial problem of having to defend territory that by its very nature wants to be in flux. Natural shores are continuously changing, with the ebb and flow of the tides, shifting of sands, and shaping through wind and storms. If we consider a shore-area that is a living and ever-changing environment, rather than a static boundary, we can begin to create new forms of coastal architecture and infrastructure that both embrace and collaborate with intertidal liminality. In the case of Singapore, a potential change in strategy from traditional land reclamation to one that adopts a nature-based approach could offer solutions for coastal development that go beyond "green" and prioritize the "more-than-human" world.

A History of Reclamation

The rapid rise of Singapore into an ultra-modern city with world-class architecture and parks is nothing short of spectacular. However, much of this new construction has been built on what was once part of the sea. Land reclamation and the transformation of the natural environment into an ordered landscape have been core to Singapore's national identity. Reclamation efforts began 200 years ago by the British colonial government and continued with the massive land building from the 1960's onward, greatly expanding the island footprint. The creation of new land has been an ongoing strategy by the government to accommodate a growing population, promote commercial development, and attract foreign investment. There has been a persistent approach of pragmatism in Singapore regarding the continued necessity of land reclamation, despite growing opposition from environmentalists, increasing difficulty in obtaining sand, and threats posed by a changing climate. The history of land reclamation in Singapore offers insights into the roots and extent of the country's coastal transformation.

When the British founded the colony of Singapore in 1819, they were attracted by the island's ideal geographical location for trade and deep-water ports, but the hills, rivers, and swamps were seen as drawbacks that needed to be remedied. Early coastal reclamation projects involved a mixture of practicality to improve the growth of commerce and efforts to develop a sanitized "colonial landscape" that established order and power.[9] The first land reclamation project commenced only a few years after the founding of the colony by Sir Stamford Raffles, near the mouth of the Singapore River. The southern bank of this important river estuary contained an extensive mangrove forest that reached several kilometers inland. At high tide the water rose as much as 10 feet, creating a flooded lagoon and at low tide became a muddy swamp.[10] This presented less than ideal features for what was to become the island's major business center. Raffles had the southern shore filled in using "Sua Kia Deng" or "Top of the tiny hill" located at the mouth of the river.[11] This was the beginning of a long history in Singapore of shoreline transformation through land reclamation over the next two centuries.

Between 1879 and 1897, the colonial government completed the first major reclamation project to redraw the shoreline along Telok Ayer.[12] The nearby hills of Mount Wallich, Mount Palmer and Mount Erskine were leveled to use as fill, extending the waterfront seaward with a 42-acre tract. The new land removed a fishing village located near the beach and covered a rocky shore at the base of Mt. Palmer with tidal pools that were places of recreation for local children.[13] The Thian Hock Keng Temple, built in 1839, was once situated on the shore of Telok Ayer. It served as a point of arrival for the Hokkien Chinese community, where recent immigrants would give thanks to the sea goddess Mazu for the safe passage. After land reclamation in the 19th and 20th centuries, the temple is now 2 kilometers from the shoreline.

In the 1930's, a large reclamation project was begun to fill in the Kallang Basin for the construction of an airport. It involved the reclamation of 339 acres of mangrove swamp that was considered "the worst mosquito-infested land on the island."[14] The Kallang Basin also happened to be the home to numerous Orang Laut villages located at the mouth of the Kallang and Rocher Rivers, adjacent to the mangrove forest and in the mud flats of the basin. Orang Laut, meaning "sea people," are the indigenous people of Singapore. Their livelihood and culture were intrinsically connected to the sea through trade and fishing. They built their houses on wooden stilts high above the water or lived in floating boat houses, adapting to the shifting shore that would soon be filled with their forced removal.

With the independence of Singapore in 1965, there was a rapid acceleration of land building, which was considered vital to its plans for national development. While the colonial government added 300 hectares through land reclamation over nearly 150 years, postcolonial Singapore would add 13,800 hectares in the subsequent three decades.[15] The "Great Reclamation" began in 1966 and involved a massive land building project on the East Coast led by the Housing and Development Board (HDB). The East Coast, known for its beautiful natural beaches, was also home to many fishing villages or "kampongs" that were removed to make way for mid and high-rise housing development. Residents recall how during the reclamation process the "whole seacoast turned into an orange mud" as the nearby Red Cliff and dredged material from the ocean were combined as fill material.[16] Others have described the loss of the pristine beaches that they once played on as the sea was "pushed away" and "ebbed further and further away."[17]

The most high-profile reclamation development is Marina Bay. Started in 1971, the 360-hectare fill development now includes financial and shopping centers, the iconic architecture of the Marina Bay Sands hotel, and Gardens by the Bay. The Garden boasts 18 Supertrees that rise "50 meters above the ground," "provide shade in the day," "are sustainable vertical gardens housing over 162,900 plants and 200 species," and "harvest solar energy."[18] In Nature's Colony, author Timothy Barnard states that the "Supertrees have become iconic symbols of the Singaporean landscape, making the complex an overwhelming metaphor of the relationship between manufactured nature and power in the small nation-state."[19] The irony is not lost that these so-called Supertrees and much of the land reclamation efforts undertaken in Singapore have replaced the real "supertrees," actual trees, such as those found in the mangrove forest, which provide nearly all of the same benefits and more.

In recent years, Singapore has been facing a shortage of sand, a resource vital in its efforts to create new land through reclamation. The United Nations Environment Program found Singapore to be the largest global importer of sand in 2014.[20] In 2010, it imported nearly 15 million tons of sand.[21] Importing sand has become more difficult over the past two decades as Malaysia, Indonesia, Vietnam, and Cambodia have all instituted bans on the export of sand to Singapore.[22] Starting in 2016, Singapore began testing out a different method of land reclamation using the Dutch polder system that reduces the amount of sand needed. A prototype is currently being constructed on the island of Pulau Tekong, which will expand the land mass by 800 hectares for a future military training base.[23]

Unlike previous land reclamation projects in Singapore, the polder system reduces the amount of sand needed for fill, by constructing a dike, along with a stormwater retention area and a network of drains, canals, and pumps. The reduction of sand needed for impoldering is especially enticing to Singapore for the savings on upfront construction costs. With a dike, buildings can be constructed below sea level. The polder area on Tekong Island will be "protected from the sea by a dike measuring 10 kilometers long, up to 15 meters wide at its crest, and will stand at about 6 meters above sea level."[24] However, by building below sea level, development will be vulnerable to any breaches of the dike wall. If history tells us anything, it's not a question of whether a dike wall will fail, but when and how badly. Additionally, polder development is dependent on pumps to remove the terrestrial water from storm events and like the dike require continuous maintenance. The costs of Singapore's extensive history of reclamation raise further questions about what has been lost culturally and ecologically, including whether the Nature in City framework can be applied to future waterfront development.

Coastal Ecology Lost

Over the past two centuries, land reclamation has covered the majority of Singapore's original coastline. It is estimated that 95 percent of the historic mangrove forests were destroyed due to "reclamation and other human activities."[25] "Losing mangrove forests not only contributes to the rapid loss of biodiversity and ecosystem function but can also negatively impact human livelihoods and the provision of ecosystem services.[26] Mangrove forests are one of the world's most productive types of wetland, providing critical habitat for numerous terrestrial, estuarine, and marine organisms, including as a food source and nursery for fish species.[27] Ecosystem services for humans include protecting coasts from erosion, tidal surge during periodic storms, and carbon sequestration. There is increasing interest in Singapore about the potential of "blue carbon" ecosystems for carbon credits to finance and support the preservation and expansion of mangrove forests.

With land reclamation covering nearly all the southern shore, Singapore not only lost its mangrove forests in the estuaries, but also its natural rocky shores, sandy beaches, and seagrass habitats.[28] In rocky shores, one might find the Pacific turban snail attached to rocks, or bristle worms and reef worms burrowing in the sand. It would have been common to see bubbler and ghost crabs walking across the sandy beaches. The seagrass meadows contained rich marine biodiversity, functioning as nurseries for juvenile animals such as crabs, shrimps, and fishes and habitat for sea stars, sea horses, and sea urchins.[29]

A Garden City

While Singapore has been filling in the coastal habitat through reclamation, it has been equally busy transforming the urban environment into a garden. The "Garden City" vision was introduced in 1967 by Prime Minister Lee Kuan Yew, father of the current Prime Minister, to create lush greenery throughout the city by planting trees and establishing parks.[30] By 2014, the city-state had planted 1.4 million trees, created 330 parks, and established 9,707 hectares of green space.[31] Examples of the prolific urban green space in Singapore abound, including the famous Gardens by the Bay and Jewel at Changi Airport. The reclamation of the East Coast resulted in not only housing development, but an extensive park and an artificial beach for public recreation. It's notable that very few Singaporeans enter the water, and many describe the beach sand as "low quality" in comparison to natural beaches in surrounding countries.

A growing interest in biophilic design in Singapore has led to increased development that integrates landscape into building design, such as courtyards, green roofs, and facades. Examples include the terraced roof gardens of Park Royal Hotel and Oasia Hotel, a 27-story mixed-use hotel and office with 21 species on its facade, both designed by local WOHA Architects. The newly constructed CapitaSpring boasts a spiraling public park that wraps around the 17th to 20th floors, in addition to the country's highest rooftop urban farm at 280 meters tall. What these constructions have in common is the incorporation of plants into the structure to brand and sell the building. With the primary goal of giving people pleasure, the architectural landscapes include plants selected specifically for their ornamental purposes.

The green spaces in cities can provide many benefits, such as reducing urban heat island effects, cleaning air and water, and flood control. These are often called ecosystem services, related to the various benefits that the natural environment provides humans. What is discussed less often is the gap in biodiversity metrics between native ecosystems that have been removed for urban development and a curated garden environment as replacement. While urban green spaces in the parks, tree-lined streets, and biophilic buildings have certainly earned Singapore a deserved reputation as the Garden City, it has primarily been for the benefit of people rather than the environment. It has been less about co-existing with nature and more about controlling it. This paradigm is especially evident on the waterfront, where tidal ecosystems have been replaced by artificial beaches and parks with little biodiversity value.

A City in Nature

In March 2020, Singapore's NParks introduced the "City in Nature" vision. Five key strategies are identified for the City in Nature 2030 plan: grow nature park networks, naturalize gardens and parks, restore nature into urban areas, connect green spaces, and enhance animal management.[32] The aim of growing nature parks networks is to safeguard the four nature reserves by keeping them separate and buffered from urban development. To naturalize gardens and parks, NParks refers to "intensifying nature" in gardens and parks to support more biodiversity and "immersive experiences in nature" for health and well-being. Examples include the use of nature-based solutions to transform concrete canals into naturalized rivers at Bishan-Ang Mo Kio Park and Jurong Lake Gardens. The goal of NParks to "restore nature into urban areas" appears aligned with the Garden City approach to plant more trees (170,000) and add skyrise greenery (200-hectares) by 2030.[33]

While the City in Nature plan is promising, it remains to be seen how much this new approach will differ from creating a Garden City and if it will truly reintroduce nature into the city. A way to consider the difference between incorporating nature versus garden into the city would be the extent to which control is yielded. When NParks refers to "naturalizing" gardens and parks, this could refer to giving up a certain degree of control in how it evolves over time and is shaped by natural processes of water, sedimentation, and plant growth. This is juxtaposed with the constant maintenance and curation that shape gardens and parks. NParks's description of "intensifying nature" may refer to both a "rewilding" of the green space and the increased biodiversity that can accompany the transformation. This distinction becomes particularly interesting when the implications of a City in Nature approach to development along Singapore's urbanized waterfront are considered.

To combat climate change and safeguard coastal areas from storm surge and sea-level rise, the City in Nature plan outlines an effort to restore mangrove forests in parks and along Singapore's coastline. However, no examples are given and the majority of coastal areas in the country that have been naturalized and restored to a native tidal ecology are outside of the urbanized areas. While lacking past examples, the City in Nature goals offer an opportunity to consider how this framework may be applied to future shoreline development that embraces natural features and processes of tidal ecosystems as an integral part of the design. *By collaborating with nature, how can future waterfront development accommodate growth and prioritize coastal biodiversity along with climate change resilience? What might it be like to walk through a mangrove forest on the way to work? Or watch otters swim amongst an aquatic seagrass meadow from your living room window?*

FIGURE 5.3   Boardwalk over the Chek Jawa Wetlands in Pulau Ubin

A Walk in the Wetlands

Since the pandemic there has been an increased interest among Singaporeans in exploring the natural areas of their country. The use of greenways and parks for exercise and hiking trails in the forest has drawn more visitors and attention to urban nature. According to NParks, this was particularly the case for Singapore's parks with more natural settings. It states that this demonstrates "how crucial nature is in providing benefits to physical and mental wellbeing as we continue to transform Singapore into a City in Nature."[34] A short boat ride from Changi Point is Pulau Ubin, Singapore's third largest island. It is one of the country's largest and best-preserved natural areas, hosting an astonishing variety of flora and fauna. The NParks website entices visitors with the promise that they will "be transported back in time to 1960s Singapore as you embark on a trip to Pulau Ubin."

On the eastern shore of the island are the Chek Jawa Wetlands that include six major ecosystems: sandy beach, rocky beach, seagrass lagoon, coral rubble, mangroves, and coastal forest.[35] A boardwalk stretches out into the water from the rocky shore, giving a vantage of the coastal forest where you might smell the sweet fragrance of the Penaga Laut tree. At low tide, Fiddler Crabs emerge from their burrows to feed, and the exposed coral below reveals a thriving marine life with anemones, sea cucumbers, and sea stars. Rounding the headland, a sandbank acts as a protective arm for a lagoon that is home to a variety of crabs, snails, and sea stars, while providing a rich feeding habitat to shorebirds like the Lesser Sand Plover and Common Redshank that winter on the coastal mudflats. Within the lagoon, sheltered between the sandbar and shoreline, is a seagrass meadow where otters often patrol the waters. Known as the nursery of the sea, the seagrass provides vital habitat for crabs, prawns, and fish to feed and spawn. As the boardwalk turns back into the shoreline, it enters a mangrove forest with its tangle of roots that reach down into the sediment, acting as a natural buffer against erosion, while branches stretch toward the sky to supply a protective canopy from the sun and a playground for Long-tailed Macaque (Figure 5.3). Further into the swampy forest is a sculpted landscape of large volcano-shaped mounds of mud, the architectural creation of the Mud Lobster. Its raised mound acts as a "condo," attracting an assortment of dwellers, including snakes, crabs, and Malayan Water Monitor lizards looking for a meal.[36]

The boardwalk leads deeper into the forest to the 20-meter-tall Jejawi Tower, unveiling an expansive view of the Chek Jawa Wetlands and Pulau Ubin. Also seen from this height are the polder-based land reclamation project underway in the nearby island of Pulau Tekong and barges filled with sand regularly passing through the strait. This view displays a juxtaposition between the past, present, and potential future of Singapore's shore. *Questions arise about the relationship between urban development and the sea. Will it continue to be a man-made battle between reclaiming new land from the sea and land being reclaimed by sea-level rise? Could a natural shoreline like Pulau Ubin be integrated into future development along parts of Singapore's urbanized shore?*

Embracing Water

In 2009, MoMA and PS1 Contemporary Art Center joined to create an "architects-in-residence" program to re-envision the coastlines of New York and New Jersey with proposals for sea-level rise adaptation that utilize "soft" infrastructure with a focus on ecological needs. The resulting 2010 exhibition *Rising Currents* was revolutionary for addressing coastal climate change resilience with designs that actively questioned the hard shoreline edge, creating instead speculative visions that embrace water and blur the distinction between land and seascape. The work also seemed prophetic after Hurricane Sandy left a wake of devastation in 2012 leading many to question why strategies in *Rising Currents* had not been previously implemented.

As a result, an even bigger sea-level rise competition was created in New York called *Rebuild by Design Hurricane Sandy Design Competition* to devise innovative approaches for future resilience. *Rebuild by Design* was developed in partnership between the U.S. Housing and Urban Development agency, non-profit and philanthropic partners, and local community and planning organizations. The competition was unique in its scope and approach, bringing together multidisciplinary design teams with community and local government stakeholders, resulting in seven funded projects to develop prototypes.

The subsequent research and designs from these competitions have created a new 21st-century toolkit on how to approach issues of climate change adaptation and resilience not previously examined. Many of the designs focus on a cohabitation of tidal ecologies along the urban waterfront and in some instances a collaboration. SCAPE, a landscape architecture firm, proposed "Oystertecture," a design that utilizes structures built underwater to support the restoration of oyster beds that once existed throughout New York Harbor and could double as ecological habitat and wave attenuation to protect cities against storm surge. In the proposal by architecture firm Lewis Tsurumaki Lewis, a low-lying site built on reclaimed land is designed not to stop flooding, but rather to build development that can co-exist with the gradual rise in sea waters and interface with the intertidal zone.

A Postcard from the Future

Imagine now a future Singapore that has embraced an adaptable waterfront that blurs the distinction between land and sea, where architecture co-exists with mangrove forests and intertidal habitat. Utilizing the City in Nature framework, areas built on reclaimed land have incorporated soft infrastructure, replacing seawalls with a naturalized shore, thus allowing tidal water into the urban environment. Sea-level rise is no longer an issue as the city-state has incorporated architectural strategies that are designed to accommodate water such as construction on piers, floating buildings, and a cut/fill approach that raises the ground level in one area while leaving an adjacent area lower for sea infiltration and stormwater management. There is less reliance on barriers to protect against a rising sea or imported sand to create new land, as neither solution is necessary for future resilience development.

In this scenario, intertidal biodiversity has flourished in cohabitation with the urban waterfront. The reintroduction of mangrove forests into the built environment provides effective climate mitigation in the form of carbon sequestration and resilience by protecting development from the more severe effects of storm surge. The mangroves in turn also benefit from human-made structures. They are sheltered by barriers that reduce the constant wave attenuation, which can lead to soil erosion from the shipping lane. Additionally, supplemental sediment is added for the trees to adapt to a rising sea that might outpace natural soil accretion. The processes of sediment and sea currents have once again established natural beaches along the shore, along with seagrass lagoons that offer both vital marine habitat and a place for residents to learn about the local aquatic species.

A pessimistic or perhaps pragmatic response to this scenario might be that Singaporeans may not want to embrace the sea and integrate new modes of building that can co-exist with water. This could be attributed to the perception that land developed through reclamation and protected with hard engineered barriers is more solid, secure, and marketable. However, to achieve goals of climate change resilience, creating sustainable development, and protecting biodiversity, new approaches will need to be tested and implemented. There are numerous soft infrastructure approaches to waterfront development that can be pulled from the *Rising Currents* and *Rebuild by Design* competitions. A similar competition based in Singapore would likely be beneficial to build place-based knowledge that focuses on the local ecology, community, and history of the island nation.

Singapore, like coastal cities around the world, has been designed to delineate a clear distinction between land and sea. Sea-level rise presents an opportunity to reconnect the built and natural environment with climate change resilient development that is not dependent on barriers, but rather integrates water and ecological systems. To truly embrace nature in the city is to give up some level of control to natural processes. It requires urban design that allows for change, for tides to rise and fall, mudflats to emerge, wetlands to establish, sandy beaches to shift, migratory birds to come and go, and storms to continually reshape a shore that refuses to be distinguished as a line. By taking a transpecies approach to designing cities, we are acknowledging our interconnection with and interdependence on the natural systems that support life on this planet, including our own.

Acknowledgement

A special thanks to Richard Bender, Joyce Hwang, and the Earth Observatory of Singapore.

Notes

1. "Green Plan 2030," SG Green Plan, Singapore Government Agency, last updated May 4, 2023, https://www.greenplan.gov.sg/.
2. Ng Huiwen, "National Day Rally 2019: 8 Things to Know about PM Lee Hsien Loong's Speech," *Straits Times*, October 26, 2019, https://www.straitstimes.com/politics/national-day-rally-2019-8-things-to-know-about-pm-lee-hsien-loongs-speech.
3. Huiwen.
4. "Sea Level Rise," PUB, Singapore's National Water Agency, last modified October 13, 2022, https://www.pub.gov.sg/Pages/sealevelrise.aspx.
5. FEMA, *Building Community Resilience with Nature-Based Solutions: A Guide for Local Communities* (Washington, DC: FEMA, 2021), https://www.fema.gov/sites/default/files/documents/fema_riskmap-nature-based-solutions-guide_2021.pdf.
6. IPCC, *Climate Change 2022: Impacts, Adaptation, and Vulnerability*, Hans-Otto Pörtner et al., eds. (New York: Cambridge University Press, 2022), 4, https://doi.org/10.1017/9781009325844.
7. IPCC, 380.
8. Matthew Palmer, Kathleen McInnes, and Mohar Chattopadhyay, "Key Factors for Sea Level Rise in the Singapore Region," in *Singapore 2nd National Climate Change Study*, Phase 1 (Singapore, 2015), 2, https://doi.org/10.13140/RG.2.1.1875.9449.
9. Miles Alexander Powell, "Singapore's Lost Coast: Land Reclamation, National Development and the Erasure of Human and Ecological Communities, 1822–Present," *Environment and History* 27, no. 4 (2021): 641, https://doi.org/10.3197/096734019X15631846928710; Brenda Yeoh, *Contesting Space in Colonial Singapore: Power Relations and the Urban Built Environment* (Singapore: NUS Press, 2003).
10. Charles Buckley, *An Anecdotal History of Old Times in Singapore* (Singapore: Fraser & Neave, 1902), 1:75.
11. "Lost Waterfronts," Channel News Asia, uploaded February 13, 2022, Singapore, video, 47:34, https://www.channelnewsasia.com/watch/lost-waterfronts.
12. Tin Seng Lim, "Land from Sand: Singapore's Reclamation Story," *Biblioasia* 13, no. 1 (04 April 2017), https://biblioasia.nlb.gov.sg/vol-13/issue-1/apr-jun-2017/land-from-sand#fn:1.
13. "Beach at the Foot of Mount Palmer," Roots, last modified April 2, 2021, https://www.roots.gov.sg/Collection-Landing/listing/1142729.
14. "New Airport is On Site of Former Swamp," *Straits Times*, June 12, 1937, 10. National Library Board Singapore.
15. Powell, "Singapore's Lost Coast," 12–13.
16. "Lost Waterfronts."
17. William Jamieson, "There's Sand in my Infinity Pool: Land Reclamation and the Rewriting of Singapore," *GeoHumanities* 3, no. 2 (21 March 2017): 399, https://doi.org/10.1080/2373566X.2017.1279021.
18. "Super Tree Grove," Gardens by the Bay, accessed May 14, 2023. https://www.gardensbythebay.com.sg/en/things-to-do/attractions/supertree-grove.html.
19. Timothy Barnard, *Nature's Colony: Empire, Nation and Environment in the Singapore Botanic Gardens* (Singapore: NUS Press, 2017), 257.
20. "Such Quantities of Sand; Banyan," *The Economist*, February 28, 2015.
21. Denis Gray, "Cambodia Sells Sand; Environment Ravaged," *Asian Reporter* 21, no. 17 (2011); "Land Reclamation in Singapore," Wikipedia, last modified February 10, 2023, https://en.wikipedia.org/wiki/Land_reclamation_in_Singapore.
22. Lindsay Murdoch, "Sand Wars: Singapore's Growth Comes at the Environmental Expense of its Neighbours," *Sydney Morning Herald*, February 26, 2016.
23. Murdoch.
24. "Polder Development at Pulau Tekong," NUS Deltares, accessed May 14, 2023, https://nusdeltares.info/projects/project-2/.
25. Powell, "Singapore's Lost Coast," 23.

26 Beth Polidoro et al., "The Loss of Species: Mangrove Extinction Risk and Geographic Areas of Global Concern," *PLOS ONE* 5, no. 4 (2010), https://doi.org/10.1371/journal.pone.0010095.
27 "Mangroves," Singapore National Parks, last updated January 2023, https://www.nparks.gov.sg/biodiversity/our-ecosystems/coastal-and-marine/mangroves.
28 Powell, "Singapore's Lost Coast," 22.
29 "Intertidal," Singapore National Parks, last updated January 2023, https://www.nparks.gov.sg/biodiversity/our-ecosystems/coastal-and-marine/intertidal.
30 Kuan Yew Lee, *From Third World to First: The Singapore Story, 1965-2000*, vol. 2, *Memoirs of Lew Kuan Yew* (Singapore: Singapore Press Holdings, 2008), 188.
31 Public Works Department, *Annual Report* (Singapore: Public Works Department, 1975), 52. National Library Board of Singapore.
32 "Singapore, Our City in Nature," Singapore National Parks, last updated January 2023, https://www.nparks.gov.sg/about-us/city-in-nature.
33 "Singapore, our City in Nature."
34 Kenneth Er, "Transforming Singapore into a City in Nature," *Urban Solutions*, no. 19 (June 2021): 68–77, https://www.clc.gov.sg/docs/default-source/urban-solutions/urbsol19pdf/09_essay_transforming-singapore-into-a-city-in-nature.pdf.
35 "Singapore, Our City in Nature," National Parks Singapore.
36 "Mud Lobster," Wild Singapore, last updated March 2020, http://www.wildsingapore.com/wildfacts/crustacea/othercrust/lobster/thalassina.htm.

6
FLOATING-WITH

Buoyant Ecologies of Collaboration and Solidarity[1]

Adam Marcus, Margaret Ikeda, and Evan Jones

Oceanic Ambitions

The ocean has long captivated the architectural imagination. As a frontier, it both feeds and limits the human impulse for expansion and growth. As a site for architectural speculation and innovation, it has inspired visionary works as diverse as the vast infrastructural efforts of Dutch land reclamation in the North Sea, Buckminster Fuller's tetrahedral floating Triton City, Ant Farm's Dolphin Embassy for human and dolphin cohabitation, and Aldo Rossi's playful Teatro del Mondo, a floating theater for Venice. These works, although incredibly divergent in scope, style, and purpose, share a communitarian ambition—an understanding of the ocean as a commons for collective living, enjoyment, or even interspecies socialization. More recently, however, the ocean has become a site of architectural pursuits far less collective in nature, as a small yet well-funded cadre of libertarian philosophers and activists have turned their gaze seaward in search of a floating neoliberal utopia.

This movement, loosely affiliated under the moniker "seasteading," melds the back-to-the-land escapism that inspired many architectural experiments in the 1960s with an extremist neoliberal politics that values individualism above all else. Funded largely by tech entrepreneurs, the seasteaders advocate for independent floating settlements to be constructed outside of territorial waters, thereby avoiding the obligations that come with collective forms of governance, such as regulation and taxation. The Seasteading Institute, a San Francisco-based organization that has received significant funding from venture capitalist and avid libertarian Peter Thiel, has become the movement's intellectual clearinghouse, providing a platform for theoretical writings, glossy PR initiatives, and a burgeoning community of startup ventures seeking to prototype floating cities. The movement's ideology is best captured in the book *Seasteading: How Floating Nations Will Restore the Environment, Enrich the Poor, Cure the Sick, and Liberate Humanity from Politicians*, published in 2017 by Joe Quirk and Institute founder Patri Friedman, grandson of neoliberal economist Milton Friedman. Quirk and Friedman's treatise weaves threads of libertarian economics, technological innovation, and environmentalist rhetoric into an argument for what they call the "colonization of the oceans."[2]

This ethos has inspired a number of architectural endeavors, most notably Oceanix City, a project designed by Danish firm BIG for startup Oceanix and developed with support from the United Nations Human Settlements Programme (UN-Habitat). This is a fitting match, as BIG founder Bjarke Ingels's philosophy of "hedonistic sustainability"[3] maps flawlessly onto the ambitions of the seasteaders to shroud the libertarian motivations of their project in a gloss of tech and ecology.[4] The project, rendered in BIG's characteristic style to resemble a kind of floating luxury resort, promises a vision of self-sufficiency, closed-loop food and water production, waste recycling, and even coral reef restoration. But this rhetoric belies the project's larger motivations: leveraging the language of sustainability and climate vulnerability to justify an escapist, privatized haven.

A Counter-Model: Collaboration and Solidarity

While seasteaders view the ocean as a colonial site for unfettered individualism and capitalism, the Buoyant Ecologies project led by the Architectural Ecologies Lab at California College of the Arts understands the ocean as an environment of dynamic flows, constant change, and collaboration across species. It explores new models of equilibrium and balance within this fluid medium. A sectional understanding of buoyancy informs the design of stable floating structures, and, more broadly, the project aligns the interests of humans and more-than-humans towards goals of mutual betterment and coexistence. As the practical considerations for this work require knowledge and expertise from multiple disciplines, the project has cultivated a collaborative working methodology with marine biologists, community organization, regulatory agencies, and fabricators.

The project began as a speculative exercise, through a series of architectural design studios in which students explored the possibilities of floating buildings for the San Francisco Bay. The design work built upon a hypothesis about ecological performance: the notion that the underside of floating structures could be designed to benefit the surrounding marine ecosystem. In various ways, the student work incorporated the premise of an "ecologically optimized substrate": an underwater growing medium that is able to promote diverse habitats for marine invertebrates based on the customization and variation of surface geometry. In testing this hypothesis, the research team developed workflows that merge techniques of design computation and digital fabrication with expertise from marine biology to develop material strategies that relate geometric difference to ecological performance.

Central to this early work was an effort to identify opportunities for mutualism: scenarios in which robust habitats for marine invertebrates provide ancillary benefit to humans. The research challenges conventional notions of "biofouling"—the unwanted accumulation of marine life on the underside of floating structures—and instead proposes controlled upside-down habitats as an ecological resource. Water flowing along this underwater landscape brings plankton and other nutrients into these "fish apartments," helping to promote ecological diversity and support biological growth that in large quantities can help attenuate wave action and reduce shoreline erosion, thereby helping to protect communities vulnerable to sea level rise. In this regard, the project explores how innovations at the micro scale of material substrates can initiate vectors of change that will have impacts at the macro, ecosystemic scale of coastal resilience.

The early work integrated critical ecological expertise from the Benthic Lab at Moss Landing Marine Laboratories and fabrication expertise from fiber-reinforced polymer composites manufacturer Kreysler & Associates, both of which facilitated the design and construction of nearly two dozen ecological substrate prototypes in the project's early stages. This interdisciplinary collaboration has enabled a positive feedback loop between the speculative thinking of architecture students, the empirical knowledge of scientists, and the fabrication know-how of industrial fabricators. Each has inspired the others to rethink the limitations and potentials of their respective discipline, with the collaborative workflow yielding outcomes that would otherwise not have been possible if pursued within a singular disciplinary model

Prototyping Regulatory Change

The proof-of-concept success of the initial substrate tests informed the design and construction of the Buoyant Ecologies Float Lab, a larger prototype launched in San Francisco Bay in 2019 that serves as a field research platform for scaling up this ecological research (Figure 6.1). The Float Lab is bean-shaped in plan, roughly the size of a small automobile, which allows it to be transported and deployed with relative ease. The prototype consists of two identical parts that form the top and bottom, like a clamshell. The geometry incorporates surface variation at two different scales of habitats: two larger "mountains" create a valley in the center of the structure, while a finer grain of surface texture is distributed across the hull. The structure also includes attachment points on the underside that are used to suspend additional substrate prototypes testing how different materials and geometries can perform as marine habitats (Figure 6.2).

As a first-of-its-kind prototype to be proposed for deployment in San Francisco Bay, the Float Lab had no regulatory precedent and required a lengthy permitting process and review by several agencies. The Port of Oakland provided the site—a mooring located in Oakland's

FIGURE 6.1 View of Buoyant Ecologies Float Lab

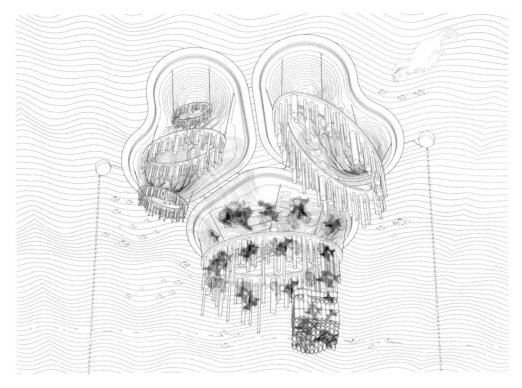

FIGURE 6.2 Fish-eye view of cluster of floating breakwaters

Middle Harbor, a restored wetland nestled within the maritime complex—which falls under the regulatory jurisdiction of the state San Francisco Bay Conservation and Development Commission (BCDC) and the federal U.S. Army Corps of Engineers. As a stationary object designed to prototype methods of increasing biodiversity and support scientific monitoring, the Float Lab did not fall into standard categories of a dock, pier, buoy, or vessel. Ultimately classified as a "floating research platform," the Float Lab was launched into the Bay in September 2019 by the Port of Oakland's engineering team, towed to its mooring at the edge of Middle Harbor, and anchored in place, where it continues to serve as a monitoring platform for material substrate experiments.

While some may consider the permitting process mundane relative to the design and ecological dimensions of this research, the regulatory approval is in many respects one of the project's most significant accomplishments to date. The San Francisco Bay regulatory apparatus, instituted in the 1960s in reaction to incremental development that compromised much of the Bay's natural marshlands, exists primarily to protect the Bay from "fill"—structures and developments that shrink the Bay's footprint, threaten the local ecosystem, and reduce public access to the shoreline. With the prospect of sea level rise now threatening to *increase* the Bay's size, BCDC recognizes the urgency to support pilot projects like the Float Lab in developing ecologically responsible approaches to climate adaptation. While a year-long permitting process may seem excessive for such a small project, its ultimate success demonstrates how interdisciplinary, collaborative initiatives like the Buoyant Ecologies project can have unique capacities to advocate for change in regulation and governance. By collaborating with regulatory stakeholders to identify a regulatory pathway for experimental prototypes, the project provides a roadmap for future pilot projects.

Future Trajectories: Floating-With

While the Float Lab continues to provide an important platform for research, data collection, and pragmatic experience with maintaining viable floating habitats, the Buoyant Ecologies project has expanded with several initiatives intended to broaden the scope of the research and its collaborative network. These include engagement and outreach efforts with students from Oakland Unified School District as well as the Treasure Island Yacht Club, in which elementary and high school students learn about the Bay ecosystem and climate adaptation through simple design exercises inspired by the Float Lab. The team has recently deployed similar ecological substrates at the restored Crissy Marsh area of San Francisco's Presidio, where they have demonstrated promise in cultivating habitats for native Olympia oysters. And recent academic studios have collaborated with partners in Maldives to explore how the premise of ecologically productive floating structures might translate to the very different climate and context of a tropical island nation facing urgent threats of sea level rise.

Cumulatively, these efforts constitute an attempt to leverage architectural thinking to establish robust, interdisciplinary collaborations dedicated to multi-species habitats and ecological benefit. As the world looks seaward with trepidation about sea level rise and its effects on vulnerable coastal communities, this work offers an alternative to the seasteaders' extractive view of the ocean. Rather than colonizing the ocean as a private enclave or tax haven, the Buoyant Ecologies project understands the ocean as a communal and collaborative space for interaction, exchange, and mutual resilience. In the spirit of Donna Haraway's notion of *sympoiesis*— the generative processes of "making-with" and co-creation that lie at the root of ecological systems[5]—the project embraces an ethos of "floating-with": with our colleagues in other disciplines, with our students, with our regulatory agencies, and with our more-than-human kin.

Notes

1 This essay was originally published in *POOL* magazine. See: Adam Marcus, Margaret Ikeda, and Evan Jones, "Floating With: Buoyant Ecologies of Collaboration and Solidarity," *POOL* v. 7: "Float" (2022).
2 Joe Quirk and Patri Friedman, *Seasteading: How Floating Nations Will Restore the Environment, Enrich the Poor, Cure the Sick, and Liberate Humanity from Politicians* (New York: Free Press, 2017), 9.
3 Bjarke Ingels Group, ed., *Yes Is More: An Archicomic on Architectural Evolution* (Cologne: Taschen, 2009).
4 For two excellent critiques of Oceanix City, see Douglas Spencer, "Island Life: The Eco-Imaginary of Capitalism," *Log*, no. 47 (Fall 2019): 167–174 and Amandy Kolson Hurley, "Floating Cities Aren't the Answer to Climate Change," *Bloomberg CityLab*, April 19, 2019, https://www.bloomberg.com/news/articles/2019-04-10/floating-cities-won-t-save-us-from-climate-change.
5 Donna Haraway, *Staying with the Trouble: Making Kin in the Chthulucene* (Durham, NC: Duke University Press, 2016), 58.

7

UNBECOMING HUMAN

Patricia Piccinini's Bioart and Postanthropocentric Posthumanism[1]

Kate Mondloch

The Australian multimedia artist Patricia Piccinini (b. 1965) creates imaginary hybrid life forms, working across various media to stage a variety of transpecies encounters. Until now, there has been remarkable continuity in readings of her oeuvre. Aside from a few fleeting references to commodity culture or the specificity of the Australian environment, the critical reception of her work has been dominated by a focus on the nature of the viewer's affective experience with the unusual creatures depicted in her works of art.[2] To some extent this is familiar interpretive terrain for bioart criticism. Christoph Cox's account in the exhibition catalog for *Becoming Animal* is representative; he observes how "uncanny, monstrous creatures … summon in us sympathies and identifications that draw us into affective relationships with the non-human."[3] In the case of Piccinini's work, however, the appreciation of the viewer's emotional and affective experience has a characteristic affirmative flavor. The majority of critics, and indeed the artist herself, tend to theorize the spectatorial experience in generous and ostensibly progressive terms of empathy, care, vulnerability, and parenting in the face of biotech experiments gone awry.

This chapter's aim is to evaluate and ultimately expand the critical discourse surrounding bioart projects such as Piccinini's. First, however, it is valuable to recognize that the category of bioart itself has various, sometimes competing, definitions. Bioart can refer to artwork engaged with representations of medical and biological research, as well as to "disruptive" uses of biological material. However, as Eugene Thacker and others have argued, reducing bioart to works that deal with biology simply as an artistic medium is problematic. According to Thacker, the term *bioart*, "marginalizes (or niche markets) art, effectively separating it from the practices of technoscience … the notion of a 'bioart' also positions art practice as reactionary and, at best, reflective of the technosciences."[4] Claire Pentecost's capacious definition of bioart is helpful for understanding Piccinini's inclusion within the genre: "Bioart projects often engage in activities broadly understood as 'scientific' (e.g., using scientific equipment or procedures, making a hypothesis and testing it, furthering an inquiry usually considered the province of the life sciences, *or addressing a controversy or blind spot posed by the very character of the life sciences themselves*)," presumably through employing any number of artistic media.[5]

Piccinini's pivotal *We Are Family* exhibition (Australian Pavilion, Venice Biennale, 2003) is this chapter's central case study. The critical reception of the bioart objects in *We Are Family* is especially interesting because the exhibition has become a testing ground for boundary-marking definitions of what it means to be human in the age of biotechnologies. The vast majority of accounts of Piccinini's *We Are Family* describe "cutely grotesque" creatures who arouse empathy and "parental" affection in their viewers. I propose instead that Piccinini's bioart objects superficially invite and yet ultimately exceed these well-intentioned but reductive characterizations. My argument knowingly treads a thin line: I contend that while the prevailing interpretations of Piccinini's work are perceptive in many ways and generally are endorsed by the artist herself, they are nonetheless hampered by a symptomatic bias that obscures other compelling contributions of the artist's practice. This is not to insinuate that I think that the artist's intention does not matter; on the contrary, I contend that her art objects exhibit a lively agency of their own that invites us to reconsider their prevailing critical reception.

Upon closer inspection, the art objects in *We Are Family* suggest a model of artistic experience that circumvents familiar categories such as compassion and parental care. Instead, these works traffic in the *un*familiar (the un-"familial"): they gesture toward a viewing experience based on an encounter with radical and inassimilable difference—an encounter with beings with whom we have no necessary ties or affinities. The persistent feel-good judgments regarding Piccinini's artistic engagement with biotechnologies—judgments that inadvertently repress the transitory moments of negative affect that viewers experience with Piccinini's nonhuman creatures—thus reveal a deep-seated anthropocentrism that may even be read as politically reactionary. Through attentive examination of the material art objects themselves, I propose an alternative interpretation of the spectatorial experience associated with Piccinini's work: a mode of viewing suggestive of a postanthropocentric posthumanism.[6] By focusing my analysis of the artwork in *We Are Family* on the topic of anthropomorphism, I aim to clarify the ethical and theoretical stakes that are implicit in the artwork and, indeed, make it a vital form of social work. This intervention generates important gains beyond merely diversifying interpretations of the artist's practice. It allows us to recognize the ways in which bioart projects such as Piccinini's invite viewers to experience and care for nonhuman others in a positive relation based not upon familiar humancentric qualities, but upon inexhaustible and potentially destabilizing difference. In the chapter's conclusion, I suggest that this relation also may offer a convincing model for ethical accountability to humans, animals, and technological beings in everyday life.

We Are Family

Piccinini's exhibition for the Australian pavilion at the 2003 Venice Biennale offers a cogent entry into her practice insofar as its immediate popular success set the stage for her debut in the international art world and established the now ingrained terms of critique.[7] Curated by Linda Michael, the suggestively titled *We Are Family* exhibition featured a dozen or so hyperrealistic yet primarily fantastical life forms and their accoutrements installed in such a way as to suggest an expanded definition of a typical Western domicile. As detailed in Michael's essay for the show's catalog, the exhibition "converted the Australian Pavilion into a home," in which viewers were invited to visit "families of human, transgenic, and unidentifiable beings."[8]

The exhibition included six pieces—a single-channel video, a group of sculptures, and four installations. *Plasmid Region* (2003), a video described by the artist as the "'heart' of the show," depicts pulsating raw fleshy biomorphic forms in an unending three-minute loop set to a reflective soundtrack of piano music in surround sound. *Team WAF (Precautions)* (2003) consists of five high-gloss helmets custom-designed for the irregularly shaped heads of a group of presumably fictitious yet adventurous and design-savvy creatures.[9] The four installations—*Game Boys Advanced* (2002), *Still Life with Stem Cells* (2002), *Leather Landscape* (2003), and *The Young Family* (2002–3)—present a series of futuristic dioramas depicting lifelike sculptures of transgenic creatures. In keeping with the exhibition's titular conceit ("We Are Family"), each installation depicts some familial scenario revolving around children. *Game Boys Advanced* consists of two casually dressed Caucasian boys engrossed in a handheld video game, who, upon closer examination, turn out to be unnervingly prematurely aged clones. *Still Life with Stem Cells* features a young white girl amicably fondling animate puppy-sized lumps of flesh roughly the same color, pliability, and texture as her own.[10] *The Young Family* shows a nursing mother and four offspring of indeterminate species (porcine in posture and scale, human or simian in eyes and appendages) sprawling on a stylish Scandinavian bed frame. Finally, in *Leather Landscape*, shoebox-sized hybrid creatures—humanoid crossbreeds inspired by African meerkats and their young—take a variety of poses atop what looks to be a postmodern leather sectional-cum-playstructure, where one of the creatures appears to be engaged with an inquisitive human toddler dressed in pink overalls (Figure 7.1).

The exhibition's hyperrealist sculptures have generated by far the most popular and scholarly interest. These objects are worth exploring in detail, especially since similar formal and

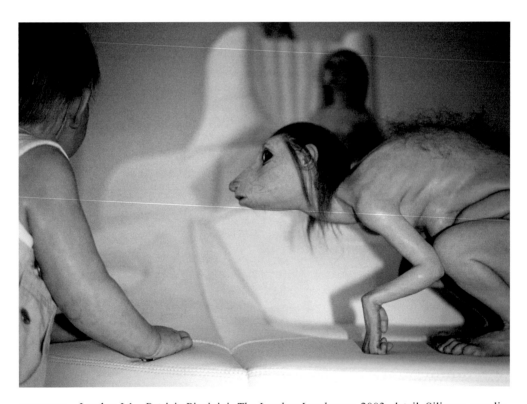

FIGURE 7.1 Leather L1—Patricia Piccinini, *The Leather Landscape*, 2003, detail. Silicone, acrylic, leather, human hair, clothing, timber. 290 cm × 175 cm × 165 cm (114 × 69 × 65 inches) (irreg.). A synthetic human toddler engages with fanciful humanoid crossbreeds inspired by African meerkats and their young.

Photograph courtesy of the artist.

thematic typologies—notably, what we might call the "adorable mutant infant" and "young humanoid children paired with displaced invented species" genres—continue to form a substantial part of Piccinini's practice.[11] The majority of the figures could be (and have been) described as endearingly monstrous. The sculptures' hyperrealism is enhanced through the use of mise-en-scène, life-size scale, genuine hair (from both animals and humans, although their application is not always species-specific), and realistic-looking flesh. (Piccinini creates the uncanny effect of lifelike skin by molding silicone over fiberglass frames.)[12] No detail is overlooked in the quest for verisimilitude: hair follicles, pores, wrinkles, and nails are so convincing that the spectator's experience might be compared to that of observing figures in a wax museum—where one knows one is in the company of faux beings yet may find oneself apologizing for backing into an inorganic starlet, Jedi, or ex-president just the same.

How do we classify the strange creatures that make their home in the Australian Pavilion? It is tempting to categorize them perfunctorily as either human or nonhuman. With the exception of the two homo sapiens girls, however, none appear to fit assuredly in those binary categories; indeed, one quickly gets the impression that this mixing of the human, animal, and technological is precisely the point. As Michael remarks of the interspecies activities staged in Piccinini's *Leather Landscape*: "The grouping conveys a sense of community, in which the child feels utterly at home…. Elsewhere a baby-sitter minds several children, enacting a role common for meerkats, whose community challenges our idea that the 'social contract' is intrinsically human. This new sculpture will assume a coincidence between the emotional and communal life of humans and animals."[13] The unusual creatures in *We Are Family* seem to trouble any secure distinction between natural and artificial creatures, and between sentient and nonsentient beings. My use of the word *beings* is deliberate; the lifelike yet discernibly nonliving sculptural entities in *We Are Family* tend to be experienced as responsive and feeling life forces owing to a range of biomorphic cues, from the obvious—eyes, limbs, flesh—to the implied—cell-like movement in *Plasmid Region*, puckered slits on the flesh lumps in *Still Life with Stem Cells*.[14] This curious implicit animism is significant because it solicits a haptic mode of viewing and contributes to the audience's self-conscious affective projections.

Piccinini's self-professed intention for *We Are Family* in many respects has produced the dominant critical response for the exhibition and much of her larger practice. It is therefore worth carefully examining Piccinini's numerous assertions about the works of art as family members. Her advocacy of the viewer's parental/familial affiliation with the eccentric techno-beings she creates is unambiguous. "There is a family inside the show," writes Piccinini, "[one] that includes us as well as these creatures that I present." On the way in which this pertains to how we should relate to the various life forms generated through biotechnology and genetic engineering in everyday life, she ruminates: "If we are family, then how does that change our attitude, how does it determine our responsibility to the creatures we create?" Ultimately she concludes that "whether you like them or you don't like them, we actually have a duty to care. We created them, so we've got to look after them."[15] Piccinini indeed routinely describes her relationship to her work as an unyielding obligation between a creator and its offspring; for example, "I can honestly say that I love my family of works…. I find them beautiful rather than grotesque, miraculous rather than freakish; but I do fear for them in a world that I see as both extraordinary and deeply problematic."[16] Even Mary Shelly's infamous Dr. Frankenstein serves as a cautionary tale about caretaking and parental obligation. Piccinini writes: "Frankenstein's mistake is that … he does not take responsibility for his creation. Having brought his creature into the world he should also be liable for its life here." Dr. Frankenstein's flaw, concludes the artist, was his failure to be a "good parent."[17]

Expanding upon the family theme in a different context, Piccinini has written: "I would never create anything I didn't love," while conceding that loving the family thus produced is not necessarily without challenges. The artist's remarks about her *Siren Mole (SO2)* (2001), a fictitious animal intended to be a commentary on genetic engineering and specifically SO1 ("Synthetic Organism 1"—the world's first fully synthetic, bacteria-like organism), is a case in point.[18] She describes her platypus-esque sculpture as "an animal that needs to be looked after, an animal in fact that cries out to be protected."[19] Due to their maladapted bodies (heavy head, pale hairless skin, and short, frail limbs), these artificial animals are in danger from both predators and the elements; indeed, "their very existence is predicated on their symbiotic relationship with us" because these artificially engineered, "pre-domesticated" creatures simply cannot make it on their own.[20] Having brought them into the world, reasons Piccinini, "we" (presumably including the audience) are obliged to nurture, love, and parent them.

Critics, for their part, echo many of Piccinini's interpretive positions. Most writers describe the viewer's experience with the artist's experimental life forms in terms of kinship, care, and child-rearing. Juliana Engberg's description, for example, affirms the role of the viewer as a parental protector: "Empathy is enhanced by the fleetingness of these creatures, or their vulnerability to the elements…. [They] worry the maternal in us."[21] Helen McDonald expands this argument: "In Piccinini's art emotion inhabits the material, everyday world and is at its most beautiful and productive when urging care and responsibility for others." Like Piccinini, McDonald associates the viewer's experience with nothing less than an ethical obligation: "[Piccinini's works] imply that the biggest challenge posed by life—for artists, scientists and other everyday citizens alike—is to care for children, nurture them, keep them safe, buffer them against unnecessary disappointment and love them in spite of their faults."[22] The persistent critical focus on human parenting and empathy is further reflected in the titles of exhibitions of Piccinini's work—*We Are Family* (2003), *(Tender) Creatures* (2007), *Hold Me Close to Your Heart* (2011), *Nearly Beloved* (2012), and the like.

It is important to recognize that even the journalistic or curatorial accounts that draw attention to the conflicted feelings viewers experience in the presence of Piccinini's unusual life forms tend to arrive at more or less the same conclusion regarding the viewer's alleged sense (or, better, *duty*) of familial care. For these critics, the viewer's initial feeling of disgust or nausea soon gives way to empathy and a sense of parental responsibility toward these life forms due to the creatures' purported cuteness and childlike vulnerability. A press release for the artist's *Relativity* exhibition (2010) puts it this way: "Often confronting yet endearingly vulnerable … [Piccinini's sculptures] assert the power of social relationships, love and communication."[23] Writing for Piccinini's *Hold Me Close to Your Heart* exhibition in 2011, curator Başak Doğa Temür confirms: "The strength of Piccinini's work evokes [a] tension through direct physical encounters as she brings the viewer face to face with hideous yet friendly creatures … the infant-like attributes of these creatures immediately evoke an inevitable sense of sympathy, care, affection, love and even an urge to cuddle and protect them."[24] Both authors contend that Piccinini's biomorphic creatures are indubitably grotesque and monstrous; at the same time, however, their infantile aspects are invoked to confirm their vulnerability and consequent lovability as our familiar and cuddly kin.

Theorizing the affective experience generated by Piccinini's practice in terms of instinctive sympathy for and guardianship of vulnerable progenies is undeniably compelling, particularly since the artist herself encourages it in her writing ("we have a duty to care") and through apparent formal cues (one need only think of the cherubic, bare-bottomed suckling mutants in *The Young Family*). In the next section, however, I will show how this seemingly agreeable interpretation willfully ignores certain qualities of the works in favor of others, thereby inhibiting appreciation of the radically estranging qualities of Piccinini's lively objects.

Are We Family?

Careful visual analysis of the works of art in *We Are Family* reveals that many of the pieces are not quite as familiar, cute, or vulnerable as dominant critical frameworks have made them out to be. To this end, I would like to invite us to linger more productively on the awkward and uncomfortable spectatorial experiences some reviewers hurriedly classify as transitory disgust or nausea. Consider *The Young Family*. While the weary nursing mother is unabashedly likable when viewed from the front, the hybrid sow creature is much less sympathetic when viewed from the rear. From this hindmost vantage point—a point of view that reveals a horse-like mane on an abject naked body that culminates in a phallic stubbed tail and eerie monkey-paw appendages—the bioengineered figure is distinctly unsettling. Once displaced from the recognizable "patient mother" genre, it is no longer clear if one should feel threatened, pitying, or neutral toward this unknown life form.

Similarly, the ostensibly agreeable Play Doh-like lumps of animate flesh in *Still Life with Stem Cells* are distinctly odd when seen from up close. The title itself suggests a pun: Are these hunks of live flesh "still" (in the sense of being immobile), or are these abject lab creatures "still" (as in nevertheless) life forms meriting our attention and respect? The creatures' hairy moles, surface veins, and wrinkles are redolent of middle-aged Caucasian humanoid skin, which generates a jarring contrast with the smooth untainted skin of the girl child cradling them. The flesh blobs' puckering body cavities—anus? mouth? gash?—further disarm the spectator in ways that complicate the dominant upbeat critical discourse. If the functional bodily cavities indicate that the flesh lumps are autonomous and ambulatory, however, do they in fact need us to care for them, as critics insist? Moreover, do "we" really make friends (family) with these strange mutant life forms as naturally and spontaneously as the artist's commentators suggest?

Linda Michael hints at the creatures' potential autonomy and lack of reciprocity in her catalog essay for *We Are Family* by identifying the possibility of a "disinterested sympathy" among viewers. Piccinini's creatures, she writes, "do not clamour for our attention—they are already in happy families. They are without the requisite 'aura of motherlessness, ostracism, and melancholy' that makes us want to 'adopt' (or more accurately, buy) consumer items." Tellingly, even despite the exhibition curator's perceptive observation, critics overwhelmingly humanize and infantilize Piccinini's creatures in their reviews of the exhibition.[25]

Plasmid Region, the exhibition's sole moving-image work, has a special place in this discussion. Despite the artist's commendation of the piece as "the heart of the show," the video is often skimmed over in reviews of the exhibition. (Revealingly, Piccinini's video works in general tend to receive cursory treatment.) *Plasmid Region*, in contrast to the at least superficially appealing and approachable nature of the other objects in the show, depicts recursive growth cycles of patently irregular, sporadically hairy, and veined tissues emitting from an unidentified pulsing organism.[26] Biomorphic fleshy orbs with uneven perforations churn at a lava-lamp form and tempo. The bulbous objects are roughly breastlike in form but their variations in scale and proliferation are doggedly perplexing. Their dark red interiors, sometimes exteriors, are suggestive of bloody wounds or growths of some kind. The title's reference to plasmids (small, circular, double-stranded DNA molecules that scientists use to clone, transfer, and manipulate genes) implies that one is looking at a kind of flesh factory, although its status as malignant or benign is pointedly unresolved.

Piccinini suggests that *Plasmid Region* represents biotechnological reproduction: "I was trying to join together biological reproduction with mechanical production, like a kind of production line that continuously pumps things out in a factory. In some ways, this symbolizes the fantasy of biotechnology, where the 'bio' and 'technology' become fused into an inseparable whole." She underlines that this mode of reproduction is characteristically uninhibited: "In some ways you might think that these floating blob things might go on to form the stem cell blobs in *Still Life with Stem Cells*, or even the helmets in *Team WAF (Precautions)*, as the forms are quite similar."[27]

Plasmid Region's interrogation of biotech fantasy is part of the artist's informal trilogy of video pieces featuring pulsing life forms—including *The Breathing Room* (2000), a video installation with sound-active floor, and *When My Baby (When My Baby)* (2005), a single-channel projection. Stella Brennan, one of the few writers to evaluate Piccinini's videos in detail, nicely summarizes the key challenge of these works. On the topic of *When My Baby (When My Baby)*, she writes: "[The video] could be read as a portrait of another life form, of another way of being. Contrariwise, it could be seen as a play on our empathy towards objects with an approximately mammalian floorplan, on our need to make our technological products look back at us, on our wish to see Men in the Moon and faces on Mars."[28] Brennan insinuates two distinct interpretations: on the one hand, the work allows viewers to experience a novel life form. Conversely, the video reflexively draws the viewer's attention to an anthropocentric impulse: the desire to construe (and arguably domesticate) unfamiliar creatures in humancentric terms.

Borrowing from Brennan, we might bridge these two analyses. Works such as *Plasmid Region* are powerful precisely because they stage encounters with "other ways of being." Moreover, they do so in such a way as to highlight the audience's promiscuous teleologies: our desire to assign human attributes and meanings to nonhuman objects and things. Put differently, being in close proximity to the nameless, faceless beings in Piccinini's moving image works makes viewers hyperaware of precisely how different—how very *non*human—these other life forms actually are. If critics have skimmed over odd details of *We Are Family*'s hyperrealistic sculptures to arrive at relatively comfortable anthropomorphic readings of the three-dimensional works, the life forms in the stubbornly ambiguous *Plasmid Region* video prove less easy to construe as kin more or less in our own image.

Plasmid Region, by featuring organic nonhuman entities that are patently hard to look at, much less to turn into amiable relatives, helps us to identify what is really at stake in the viewer's experience with the various life forms assembled in *We Are Family*. Much like Piccinini's video, the hyperrealistic sculptures, too, offer encounters with untamable difference, inviting viewers to reflexively recognize that these entities cannot simply be reduced to human-centered interactions. The staying power of prevailing interpretations of parental obligation toward our cute (if grotesque) offspring thus requires further analysis.

In the next section we will consider what this feel-good emphasis might mean and what might it say about our ongoing boundary-mapping practices between human and nonhuman beings. Donna Haraway's description of the material-discursive production of knowledge is germane here. She writes: "Bodies as objects of knowledge are material-semiotic generative nodes. Their *boundaries* materialize in social interaction…. Siting (sighting) boundaries is a risky practice."[29] As we will see in what follows, Piccinini's *We Are Family* and its critical reception not only exemplify how lively, productive, and unfixed matter can be but also underscore the promise in remapping established boundaries.

Unbecoming Humanism

On some level the unfortunate generalizations about Piccinini's practice can be attributed to hasty analysis, perhaps aided and abetted by the photocentric dissemination and reception of art objects meant to be experienced firsthand. The challenges associated with the photo-mediated reception of material and experiential works of art are of course a challenge for all art criticism, particularly in the Internet era when JPEGs are routinely circulated to represent experiential works of art.[30] This issue bears special relevance, however, in cases where the critical discourse centers around issues of the spectator's immediate affective experience with purposefully mimetic and biomorphic objects (as is the case with Piccinini's practice). The limited and necessarily incomplete photographic representations of Piccinini's work undoubtedly have contributed to a critical discourse that largely ignores the very aspects of located, fleshly embodiment that generate the works' productive moments of discomfort and "bad" affect. What is unmistakable, however, is that even as these hybrid technobeings fall outside of our familiar taxonomies, *We Are Family*'s interpreters work tirelessly to explain the artistic experience with the various "embryonic figures, babies, mothers and clones" in recognizable and nonmenacing terms. The insistence upon the vulnerability of the creatures and our need to take care of and/or parent them as family exposes a deep anthropocentric bias, a desire to remake everything in (hetero)normative humancentric terms, irrespective of competing evidence—transgenic markings, alien body parts, indecipherable countenances, and the like.

In contrast to dominant critiques, Haraway has cogently argued that Piccinini's works such as *Bodyguards/Nature's Little Helpers* (2004) exceed heteronormative concerns insofar as they provoke the question of care for intra- and interacting generations beyond mere reproduction. We will return to this proposition shortly.[31] For now, however, the apparent and unqualified promotion of the heterosexual nuclear family and humanist logic in the critical and popular reception of *We Are Family* merits careful scrutiny. However well-intentioned, the prevailing interpretive framework not only conceptualizes these entanglements of human and nonhuman entities within the well-known confines of the heterosexual nuclear family, it also betrays a paternalistic and primitivizing impulse vis-à-vis the Other: Piccinini's anatomical pariahs are unintimidating if—and only if—they are experienced as obvious subordinates. ("We" must take care of "childlike" others.)

Daniel Harris offers an explanation for the potential attraction of the "cutely grotesque" aesthetic in his *Cute, Quaint, Hungry, and Romantic: The Aesthetics of Consumerism*. He writes: "The grotesque is cute because the grotesque is pitiable … the aesthetic of cuteness creates a class of outcasts and mutants, a ready-made race of lovable inferiors."[32] Harris goes on to link this impulse to a seductive consumerism. The sanguine focus on cuteness and sentimentality in the discourse surrounding Piccinini's oeuvre has implications even beyond commercial viability in the art world, however. At a time when biotech stands to make a profit off of all that lives, and life itself is computable and malleable in dazzling and dizzying ways, the significance of the fact that Piccinini's radically different creatures are consistently described as nonthreatening, infantile, and controllable should not be discounted.

The prevailing critical reception of Piccinini's practice has effectively obscured other pressing ethical and political questions about biotech's promissory rhetoric of endless progress, enhancement, and empowerment of the human being via science by speculating almost entirely about the affirmative familial quality of our relationships to its ostensibly cutely grotesque mishaps.[33] Helen McDonald's optimistic description is exemplary: "Piccinini's art is infused with hope that these kinship ties and her affectionate characterizations will engender a sense of responsibility in today's technological world."[34] The critical propensity to understand the artworks as cute and sentimental (as "affectionate characterizations") promotes a reassuring illusion of the human subject's mastery and agency ("responsibility in today's technological world") that obscures the political and social challenges associated with developing commercial products made from biological systems. The sentimental critical focus on the alleged "cuteness" of biotech's incongruous yet inevitable accidents (asking us to say "aaawww" instead of "aaahhh!" as it were) furthers the illusion of human exceptionalism, agency, and control that the contemporary technosciences otherwise undermine.[35]

At the most fundamental level, then, the impulse among Piccinini's commentators to subordinate and sentimentalize the nonhuman creatures in *We Are Family* suggests an unintentional reappearance of a traditional humanism. This return is unfortunate because it neutralizes the potentially destabilizing difference of the otherwise enigmatic biotech bodies populating her practice by conceptualizing them in relation to a universal and all-powerful human subject. As Haraway has skillfully declared: "The discursive tie between the colonized, the enslaved, the noncitizen, and the animal—all reduced to type, all Others to rational man, and all essential to his bright constitution—is the heart of racism and flourishes, lethally, in the entrails of humanism."[36] It is interesting to reconsider curator Michael's claims for *We Are Family* with this observation in mind. According to Michael, Piccinini's works urge us to consider the "coincidence between the emotional and communal life of humans and animals."[37] Following Haraway, the seemingly progressive idea of treating other species as our kin by extending *human* rights to them could also be interpreted as yet another infelicitous endorsement of human exceptionalism—a rather disingenuous attempt to detect a shared *humanity* uniting divergent species. After all, the often-cited quality of empathy is a *human* ethical value and not necessarily a universal, transpecies one.[38]

As we have seen, understanding Piccinini's work in the anthropocentric terms of sympathetic bonding and familial obligation portends a reactionary return to the surety of human power and distinctiveness. More to the concern of the present argument, however, is the way in which this unintentional critical prejudice also misrepresents the potentially disruptive nature of the critical proximity otherwise promoted and performed by Piccinini's work. While Piccinini's commentators are correct to emphasize the viewer's highly affective experience with these different species and life forms, and to discern an implicit model of care, the supposedly inescapable reactions of parental nurturing and bonding are in fact premised upon discursively fixing the creatures within stabilized and anthropocentric frames of reference. Examining the art objects up close and with renewed attention to disruptive moments of negative affect instead exposes the discrete and perhaps even unknowable identities of *We Are Family*'s bioart objects.

Uneasy Nature (or, Toward a Postanthropocentric Posthumanism)

The viewer's experience with Piccinini's nonhuman life forms can be understood constructively as a relationship based *not* on the objects' endearing humanoid infantile qualities but, rather, on their inexhaustible otherness. Appreciating the ways in which our experiences with Piccinini's creatures fall outside of familiar and recognizable categorizations is transformative because it destabilizes the traditional centrality of the human subject. In so doing, it allows us to appreciate how the bioart works in *We Are Family* do indeed solicit a relationship based on a sense of affinity, but a sense of affinity that operates along nonanthropomorphic lines. Instead of merely encouraging us to remake others as ourselves (as prevailing accounts imply), these works of art urge us to understand human and nonhuman beings as qualitatively different and yet profoundly entwined and coconstituted.

Haraway's observation about the philosophical challenges these novel human–nonhuman interfaces propagate is apposite: "To care is wet, emotional, messy, and demanding of the best thinking one has ever done."[39] Rosi Braidotti offers one such response in defining what she calls postanthropocentric posthumanism. In *The Posthuman*, Braidotti designates this approach as "post-anthropocentric" in that it takes nonhuman "organic others" seriously, and "post-humanist" in that it eschews the erroneous sense of cognitive self-mastery associated with traditional humanism in favor of appreciating and celebrating our flexible and multiple contemporary identities. Posthumanism should not be confused with *anti*humanism, however. Braidotti is at pains to explain how this radical feminist project is by no means groundless or without an ethics: "To be posthuman does not mean to be indifferent to the humans, or to be de-humanized. On the contrary, it rather implies a new way of combining ethical values with the well-being of an enlarged sense of community, which includes one's territorial or environmental inter-connections."[40] Postanthropocentric posthumanism, then, celebrates the affirmative connections between human subjects and multiple, unfamiliar (unfamilial) others. It rejects the idea of a universal normative human subject and autonomous subjective agency, but rejoices in the humanities and the cultivation of human ethics committed to expansive definitions of community and concern.

It is here that Piccinini's art practice may serve as a potential bridge: a practical, material manifestation of the speculative projects outlined by feminist theorists such as Braidotti. When seen in their full potential, the lively objects assembled in *We Are Family* direct attention to a mode of subjectivity that eschews humanist self-centered mastery while affirming nonhuman organic others—transmuted pigs, elderly children, ambulatory tumors—in all their destabilizing difference. This openness toward alterity is profound because it occurs despite any actual shared social identity, morality, or species, and despite (indeed, perhaps even because of) transient moments of negative affect. Nicole Seymour's delineation of a queer ecological ethic of care gets to the heart of the matter. The care in question for Seymour brings us full circle to Haraway's astute identification of an intra- and interacting generational model of care in Piccinini's work. It is "a care not rooted in stable or essentialized identity categories, a care that is not just a means of solving human-specific problems, a care that does not operate out of expectation for recompense."[41] Although written in a different context, Petra Lange-Berndt suitably encapsulates the transformative potential of this postanthropocentric posthuman ethic. She writes: "If the knowing self is partial, never finished and whole, it can join with another, to see together without claiming to be another. Phenomenology insists on a macroscopic, anthropomorphic view, while to be complicit with the material means, above all, to acknowledge the non-human."[42] As we have seen, the spectator's experience with the dynamic bioart objects in *We Are Family*, upon closer inspection, does not necessarily align with the dominant interpretations; far from soliciting an anthropocentric parenting of cute inferior beings, these works of art generate an artistic experience that provocatively enacts precisely the nonhierarchical material-relational complicity with nonhuman others outlined by Braidotti, Haraway, Seymour, and others.

In conclusion, while *We Are Family* has provided our central case study thus far, it is interesting to note that the model of artistic experience as a nonessentialized nonhuman-subject-centered ethic of care can profitably be discerned in aspects of Piccinini's production beyond her 2003 exhibition. Indeed, one of the artist's early works, *Protein Lattice—Subset Red, Portrait* (1997), is especially instructive in this regard. The photograph presents a truly exceptional vision of what Braidotti's proposal for a postanthropocentric posthumanism might look like on an applied basis. Two highly groomed, fleshy, and exotic others are on display—a fashion model and a synthetic rat inspired by the famous Vacanti mouse (a laboratory animal with what looks like a human ear grown on its back)—with no clear hierarchy between human and nonhuman.[43] Practiced models, they gaze laconically in opposite directions. Neither creature looks dismayed or even particularly surprised to be together. We do not know who the star is, or who serves as the pedestal for whom. Indeed, there is a suggestive visual interpenetration between their forms. The fleshy, almond-shaped ear perched on the rat's back echoes the woman's mouth, eyes, shoulder, and breast. The woman's eye shadow is fur-colored and her entire eye from brow to lower lids is rodent-shaped. The two primary orifices (her mouth, its ear) are so near to one another and so similar in shape that they seem almost ready to merge or embrace. The hairless animal conceivably could have grown out of the model's shoulder or been affixed there through some fleshy interface (Figure 7.2).[44]

In assessing this photograph, the quintessential anthropocentrism of rat experimentation for human beauty products or organ harvesting (in the form of the presumably human ear on the rodent's back) immediately springs to mind. Indeed, this image has frequently been mobilized in the service of articles on both topics. Upon further analysis, however, the relative agency of each subject is less clear. The woman's hand appears almost clawlike due to the suggestive proximity of the animal. Human fingernails are rendered alien, and wrist wrinkles rodent-like. One of the model's fingers could easily be construed as part of the rat. (The prospect is not so outlandish if one considers that the rodent has already acquired a human ear.) It is equally conceivable that the human life form could be growing parts for the rat's benefit (some tiny rat feet between the fingers of her hand, perhaps?). Appropriating Braidotti's vocabulary, we might identify the animal–human interaction depicted in *Protein Lattice* as an ethical relation based on positive grounds of "joint projects and activities."[45] While viewers cannot know for sure what "joint project" the duo might be undertaking, they can readily recognize their alliance and mutual care. Rather than offering spectators a reassuring illusion of mastery based on a heteronormative familial bond, the photograph affirms how human subjects are coconstituted with the inassimilable difference of nonhuman others (cute, rodent-like, or otherwise).

As we have seen through detailed analysis of artworks ranging from the *Protein Lattice* photograph to the multimedia objects assembled in *We Are Family*, Piccinini's practice offers a critical perspective that has long gone unrecognized. Piccinini's nonhuman life forms are *not* simply family: instead of merely promoting parental affection in viewers, her work models a form of ethical relation in which care or a sense of entangled responsibility need not be based on anthropocentric or even biological frameworks. In this way, it may offer a practical model for conceptualizing the ethics of contemporary human–nonhuman relations, both within the institutional context of the visual arts and beyond. "Ethics," clarifies Karen Barad, "[is] not about right responses to a radically exteriorized other, but about responsibility and accountability for the lively relationalities of becoming, of which we are a part. Ethics is about mattering,

FIGURE 7.2 Red Portrait—Patricia Piccinini, *Subset Red* (Portrait), 1997. From the series *Protein Lattice*, Type C photograph. 80 cm × 80 cm (31.5 × 31.5 inches). This photographic pairing of human and nonhuman features a fashion model and a synthetic rat inspired by the famous Vacanti mouse (a laboratory animal with what looks like a human ear grown on its back).

Photograph courtesy of the artist.

about taking account of the entangled materializations of which we are part, including new configurations, new subjectivities, new possibilities."[46] Piccinini's work promotes an ethic of care stripped of human exceptionalism and standardized identities, dissociated from traditionally humanist and anthropocentric demands for reciprocity or guarantees of success; instead, it acknowledges, preserves, and affirms boundless difference and collaborative exchanges within context. To encounter these works of art in their fullness is to come face-to-face with the inexhaustible, nonreciprocal otherness of nonhuman technological beings, and, crucially, to care for them all the same.

Notes

1. This essay originally appeared in Kate Mondloch, *A Capsule Aesthetic: Feminist Materialisms and New Media Art* (University of Minnesota Press, 2018). *Copyright 2018 by the Regents of the University of Minnesota.*
2. Donna Haraway interprets Piccinini's oeuvre, following anthropologist Deborah Bird, in relationship to indigenous Australian models of care for Australia's land and peoples in her "Speculative Fabulations for Technoculture's Generations," in *(Tender) Creatures* (Vitoria-Gasteiz, Spain: Atrium Gallery, 2007). See also Deborah Bird Rose, *Reports from a Wild Country: Ethics for Decolonisation* (Sydney: University of New South Wales Press, 2004). While Haraway's essay is predominantly interested in the locational, Australian context of Piccinini's practice, I will return to her sympathetic reading of Piccinini's work in the context of a queer ethic of care in the body of the text.
3. Christoph Cox, "Of Humans, Animals, and Monsters," in *Becoming Animal: Contemporary Art in the Animal Kingdom*, curated by Nato Thompson, Massachusetts Museum of Contemporary Art (Cambridge, MA: MIT Press, 2005), 23. *Becoming Animal* featured Piccinini's sculpture *The Young Family* (2002–3), which first debuted in the *We Are Family* exhibition at the Venice Biennale in 2003.
4. Eugene Thacker, *Global Genome: Biotechnology, Politics, and Culture* (Cambridge, MA: MIT Press, 2005), 307.
5. Claire Pentecost, "Outfitting the Laboratory of the Symbolic: Toward a Critical Inventory of Bioart," in *Tactical Biopolitics: Art, Activism and Technoscience*, eds. Beatrice da Costa and Philip Kavita (Cambridge, MA: MIT Press, 2008), 110. Emphasis added.
6. Rosi Braidotti coins the term "postanthropocentric posthumanism" in *The Posthuman* (Cambridge: Polity Press, 2013). I address her analysis in detail in the later part of the chapter.
7. Indicative of the exhibition's widespread popularity, *We Are Family* was restaged in its entirety at the Hara Museum of Contemporary Art in Tokyo in December 2003, immediately following the Venice Biennale showing. Despite the fact that Piccinini is consistently written about, excerpted, and illustrated in relationship to critical debates on biotech, the artist herself does not claim any expertise or critical position on science. She describes herself as an "interested amateur" whose primary contact with advanced technologies is through a subscription to the generalist magazine *New Scientist*.
8. Linda Michael, *Patricia Piccinini: We Are Family* (Strawberry Hills, NSW: Australia Council, 2003), 10.
9. While the polished crash helmets in *Team WAF (Precautions)* initially might seem like outliers to this family dynamic, the fiberglass helmets form part of Piccinini's larger automotive-inspired production in which biomorphic automotive parts appear to consciously engage in intimate encounters. (For example, the Madonna and child Vespas in *Nest* [2006], and the infantile *Truck Babies* [1999], who gaze adoringly at their equally cute preteen guardians in a multimedia installation titled *Big Sisters* [1999].)
10. The "skin" color of the biotech creatures in *We Are Family* unmistakably depicts Caucasian flesh tones. The artist's broader oeuvre demonstrates more diversity.
11. In addition to the many exhibition catalogs cited throughout this chapter, the artist's website provides a useful illustrated overview of Piccinini's production: http://www.patriciapiccinini.net.
12. Piccinini works extensively with expert craftspeople to produce her sculptures, including hyperrealist sculptor Ron Mueck and his studio. For an account of the specific work processes of the two artists, particularly as they relate to contemporary discourses on appropriation and the managerial ethos of contemporary art, see Linda Williams, "Spectacle or Critique? Reconsidering the Meaning of Reproduction in the Work of Patricia Piccinini," *Southern Review: Communication, Politics, Culture* 37, no. 1 (2004): 76–94.
13. Michael, *Patricia Piccinini: We Are Family*, 6.
14. Bioartists Oron Catts and Ionat Zurr of the Tissue Culture and Art Project (TC&A) refer to a class of object/being they call the "Semi-Living," although, in contrast to Piccinini, their practice involves the literal use of tissue technologies as an art medium. Marcos Cruz considers the relationship between Piccinini's depictions of inanimate, yet pulsing life forms and the strange-looking lumps of autonomous flesh that prove to be somehow alive in David Cronenberg's "body horror" film *eXistenZ* (1999) in "Synthetic Neoplasms," *Architectural Design* 78, no. 6 (2008): 36–43.
15. Patricia Piccinini, public presentation at the Tokyo National University of Fine Arts and Music, Faculty of Fine Arts, Tokyo, Japan, December 8, 2003.
16. Patricia Piccinini, "Patricia Piccinini in Conversation with Alasdair Foster," *Photofile* 68 (2003): 22.
17. "Artist's statement on *SO2*," Patricia Piccinini, accessed November 10, 2015, http://www.patriciapiccinini.net.

18 She has designed and reproduced *SO2* in a number of contexts. The strange beast has appeared in photographs as the passenger in the front seat of a Holden car in *Waiting for Jennifer* (2000), as the playmate of young boys in *Social Studies* (2001), as a laboratory animal in the *Science Story* series (2002), and in three-dimensional form installed in the wombat enclosure of the Melbourne Zoo (2001).
19 Piccinini, public presentation at the Tokyo National University of Fine Arts and Music, December 8, 2003.
20 "Artist's statement on *SO2*."
21 Juliana Engberg, "Atmosphere," in *Patricia Piccinini: Atmosphere, Autosphere, Biosphere*, Juliana Engberg, Edward Colless, and Hiroo Yamagatam, eds. (Collingwood, Australia: Drome Pty Limited, 2000).
22 Helen McDonald, *Patricia Piccinini: Nearly Beloved* (Dawes Point, NSW: Piper Press, 2012), 123.
23 Media press release for *Patricia Piccinini: Relativity* (Perth, WA: Art Gallery of Western Australia, 2010), http://www.artgallery.wa.gov.au/about_us/documents/Patricia-Piccinini-media-release-2010.pdf.
24 Başak Doğa Temür, "Just Because Something Is Bad, Doesn't Mean It Isn't Good," in *Hold Me Close to Your Heart* (Istanbul: ARTER, 2011), http://www.patriciapiccinini.net/printessay.php?id=37.
25 Michael, *Patricia Piccinini: We Are Family*, 18.
26 While most reviewers do not explore the possibility, *Plasmid Region* could also be understood in relationship to the grotesque as it has played out in contemporary art in discourses on abjection and the *informe*. See, for example, Yve-Alain Bois and Rosalind E. Krauss, *Formless: A User's Guide* (New York: Zone Books, 1997).
27 Piccinini, public presentation at the Tokyo National University of Fine Arts and Music, December 8, 2003.
28 Stella Brennan, "Border Patrol," in *Patricia Piccinini: In Another Life* (Wellington, NZ: Wellington City Gallery, 2006), 6.
29 Donna Haraway, "Situated Knowledges: The Science Question in Feminism and the Privilege of Partial Perspective," *Feminist Studies* 14, no. 3 (Fall 1988): 595.
30 See, among others, Hito Steyerl, "In Defense of the Poor Image," *e-flux* 10 (November 2009), http://www.e-flux.com/journal/10/61362/in-defense-of-the-poor-image/ and Orit Gat, "Global Audiences, Zero Visitors: How to Measure the Success of Museums' Online Publishing," *Rhizome*, March 12, 2015, http://rhizome.org/editorial/2015/mar/12/global-audiences-zero-visitors/ for accounts of this and related trends in contemporary exhibition and publicity related to contemporary art.
31 Haraway, "Speculative Fabulations," 3. See also Haraway's discussion of Piccinini in *When Species Meet* (Minneapolis: University of Minnesota Press, 2008), 288–91.
32 Daniel Harris, *Cute, Quaint, Hungry, and Romantic: The Aesthetics of Consumerism* (New York: Basic Books, 2000); quoted in Michael, *Patricia Piccinini: We Are Family*, 21. I am indebted here to Michael's catalog essay in which the author proficiently details the ways in which Piccinini's oeuvre draws us "perilously close to [Harris's] mode of engagement," yet ultimately circumvents this critique, in part by asserting the "redemptive power of social values and relationships"; Michael, *Patricia Piccinini: We Are Family*, 21.
33 It is worth emphasizing that the problem is not merely with biotech, but science in the service of neoliberalism more broadly, in which twenty-first-century scientific and technological production benefits from traditional claims to truth and service to the public even while increasingly tied to narrow commercial agendas. For an introduction to these debates, see, for example, Melinda Cooper, *Life as Surplus: Biotechnology and Capitalism in the Neoliberal Era* (Seattle: University of Washington Press, 2008). On the topic of bioethics as pertinent to cultural and artistic concerns, see Joanna Zylinska, *Bioethics in the Age of New Media* (Cambridge, MA: MIT Press, 2009), especially chapter 5, "Green Bunnies and Speaking Ears: The Ethics of Bioart."
34 McDonald, *Patricia Piccinini: Nearly Beloved*, 35.
35 The culturally specific impulse to interpret Piccinini's works as tender creatures requiring our parental affection came into sharp relief in April 2006 when an image of one of the animal figures in Piccinini's sculpture *Leather Landscape* was repurposed on a Sudanese Arabic-language website. As reported by Brennan, the image was circulated in the Muslim world through newspapers, schools, and mosques, and on Islamic message boards to illustrate a cautionary tale about a girl who was transformed into an inhuman beast for abusing the Koran. Brennan, "Border Patrol," 7.
36 Haraway, *When Species Meet*, 18. Haraway's classic cyborg manifesto—which playfully celebrates the cyborg figure for its ability to transgress boundaries of gender, race, and difference and to disrupt ontological categories—is the foundational text in this regard. Donna Haraway, "A Manifesto for

Cyborgs: Science, Technology, and Socialist Feminism in the 1980s," in *Simians, Cyborgs and Women* (New York: Routledge, 1991), 149–81.
37 Michael, *Patricia Piccinini: We Are Family*, 18.
38 Rosi Braidotti stresses the need to conceptualize human/animal as constitutive of the identity of each, as opposed to anthropomorphizing animals as emblems of the universal ethical value of empathy, in *The Posthuman* (Cambridge: Polity Press, 2013), 79.
39 Haraway, "Speculative Fabulations," 14.
40 Braidotti, *The Posthuman*, 190.
41 Nicole Seymour, *Strange Natures: Futurity, Empathy, and the Queer Ecological Imagination* (Urbana: University of Illinois Press, 2013), 184. Seymour's embrace of compassion, optimism, and future generations is especially noteworthy in relationship to Lee Edelman's extended critique of reproductive futurism and environmental agendas grounded in heterosexist, proreproductive rhetoric. Lee Edelman, *No Future: Queer Theory and the Death Drive* (Durham, NC: Duke University Press, 2004). On the general interest in fluidity and indeterminacy among queer theorists as a way to demonstrate the unstable distinction between the human and nonhuman, see, for example, Noreen Giffney and Myra Hird, eds., *Queering the Non/Human* (Farnham, UK: Ashgate, 2008).
42 Petra Lange-Berndt, *Materiality (Documents of Contemporary Art)* (Cambridge, MA: MIT Press, 2015), 17. For an interesting account of how artists have engaged with nonanthropocentrism, see Katherine Behar and Emmy Mikelson, eds., *And Another Thing: Nonanthropocentrism and Art* (Brooklyn, NY: Punctum Books, 2016).
43 The "ear" on the original Vacanti mouse was actually an ear-shaped cartilage structure grown by seeding cow cartilage cells into a biodegradable ear-shaped mold and then implanted under the skin of the mouse. Note that Piccinini's rendition is not a mouse but a rat.
44 When *Protein Lattice* is installed as part of a complete photographic series, a row of video monitors plays a looped narrative beneath the nine photographs. McDonald explains that the video footage emulates the experience of playing a video game and features digitally constructed vignettes from the points of view of the mouse and the spectator, eventually collapsing the two perspectives. McDonald identifies this as a progressive empathetic exchange: "we will never know what it is to be the other, but there may be benefits in trying to imagine ourselves in the other's place;" McDonald, *Patricia Piccinini: Nearly Beloved*, 45.
45 Braidotti, *The Posthuman*, 190.
46 "Interview with Karen Barad," *New Materialism Interviews & Cartographies*, Rick Dolphijn and Iris van der Tuin, eds. (Ann Arbor, MI: Open Humanities Press, 2012), 69.

8
SALT FORMATIONS

Rosalea Monacella

The inland sea of Kati Thanda (Lake Eyre), Australia, rhythmically beats, dramatically expanding and contracting as its slow-beating heart comes to life with amazing vigour after long periods of dormancy to unearth an explosion of colour, abounding wildlife, and microbial vitality.

The desolation these great floods cause over the bare, wind-ravaged wasteland is temporarily, at any rate, as cruel and destructive a master as aridity and desiccation. But in the inner sanctum of the Dead Heart, where the Tirari and Simpson Deserts surround the lake, no one, other than a handful of stockmen, ever sees it, and the devastation, like most else in these deserts, is left alone in ravaged agony. Yet, around the ocean as blue as wild violets, smiling innocently serene, after its mad race to journey's end, there is growth so lush on the sandhills, that when it is unavoidably trampled underfoot, there rises up that earthy, pungent, juicy odour of crushed rain-soaked herbage.[1]

This account draws on the journey of three people on two separate expeditions who travelled across the immeasurable salt lakebed of Kati Thanda. In 1972, Roma and John Dulhunty were in pursuit of scientific knowledge about the Great Artesian Basin, in which Kati Thanda is located. John, a geologist, and his wife Roma, his research assistant, retraced the expedition of fellow British geologist and founding professor of the Geology Department at the University of Melbourne, John Walter Gregory. In 1901, Gregory wrote the book, *Dead Heart*, a term that he often used to describe Lake Eyre. He called it a "barren wasteland," incorrectly believing it to be "lifeless/dead" with nothing living or growing.[2]

This chronicle is constructed through multiple viewpoints and voices, namely those of Roma and John Dulhunty, John Walter Gregory and of the soil, water, and sky of Kati Thanda. It aims to reveal the unique qualities of Kati Thanda's landscape through the indeterminate act of "worlding" – an act of becoming in which the observer and the observed are an extension of, and embedded in, the lens of life that reveals a flourishing landscape, where human and non-human life is entangled through the ground – in its soil, silt, sand, salt, and the millions of organisms contained in the strata that comprise this spectacular landscape.[3]

The act of tracing, in this account, is generated from a multitude of encounters through which knowledge and information are transferred. It draws on multidisciplinary knowledge systems from the arts, biology, geology, political economics, etc. As the vast everchanging landscape of

DOI: 10.4324/9781003403494-10

Kati Thanda defies customary scales of measure, it is necessary to understand the encounters in, of, and through the landscape by disassembling the idea of measure as a set of nested scales that neatly fit within one another as a scalable cartographic endeavour.

Consequently, recording equipment, tools, and techniques that are conventionally utilised to map and scale landscapes are rendered defunct here. Alternative ways of seeing define a new system of measurement that consequently inform a geographic paradigm shift. This is a living model for a new form of "world-making" that is more attuned to the landscape, its species – both human and non-human – and the world they make.

As we journey through this endorheic water system, which retains water and does not release it into an ocean, we explore forms of measurability without projective structures of mathematical precision, scalability, or stability. Measurement is driven, instead, by the dynamics of the landscape and its associated systemic drivers.

Conventional systems of mapping and measuring landscapes have most often been driven by techniques and methods that support behaviours of ownership and extraction, with a discrete separation between the method of measurement and the landscape itself. The climate crisis has spotlighted the impact of our voracious global appetite for resources and has highlighted the manner in which mineral extraction is based on these systems and the principles that drive them.

In a similar fashion, mapping and geographic systems, such as geographic information systems, Google Earth, and the like, fulfil people's insatiable desire to pan endlessly across the virtual globe (the earth's digital twin), zooming in and out in a seemingly infinite manner. This feeds the desire – and capacity – for infinitely scaling information, leading to a false sense of precision in an abstract projected Cartesian plane of the earth. It promotes a method of worlding that is infinitely scalable and ever-expanding.

The paradox of the Kati Thanda lakebed is that it was intrinsically born out of extreme conditions of wet and dry, of boom and bust, from abundance to scarcity, all brought about by the swinging pendulum of La Niña weather patterns. As a natural meteorological feature, Kati Thanda is disorienting. It questions the usual distinctions between ground and water, human and non-human, up and down, and time and space, as the infinite horizon blurs the domains of ground and sky. A shimmering of light on the lake's salt-encrusted surface reveals an array of delicate, moist pink crystals that are slowly transforming, shifting, and gathering in the heat of the midday sun. This semi-translucent surface, with vertical undulations of only a few centimetres measured across vast horizontal distances, holds life within its substance. Dormant fish species and plant life, in stasis in this thin salt layer, wait to awaken. Artesian springs and aquifers bubble away from underneath, only expressing their full capacity when the salt crust is upturned by the spontaneous downpours that flow from north to south across this vast continent.

A fine-tuned sensing of matter, time, and the various life forms supported by the irregular emergence of the ephemeral lake is required to understand what imbues this dynamic formation and transformation. This landscape is defined by its fluvial and geological processes and operations of change at a multitude of material scales, rather than only figure and form, which serve only to reinforce issues of scalability, extraction, and colonisation. The visually striking surface of the salt plain is an expression, or sign, projecting simultaneously into the future and back into its past.[4] Formed by sub-surface systems of hydrology and geology – in concert with shifts in near-invisible flows of historic magnetic fields, wind, and light – the landscape exemplifies the condition of the thickened ground. Mapping and constructing a figure of this ground is a process of uncovering and reading in ways that are an extension of the landscape, its structures, and material processes. This landscape and the subsequent form of the entire earth are understood as a network of ecosystems in a perpetual state of service, from the microbial to the planetary, formulating an indeterminable map as the ground continues to shift.[5]

The indexing and tracing of the flows and formations of the landscape through its perpetual transformations generate a counter map that reveals a sensed landscape, where time is infused recursively in the material reality of the landscape through states of formation, from those that signify stability to sequences that are predictable and observable processes of change and those that are uncertain and instantaneous. A set of narratives can be established. Geological time, microbial life, and ethereal atmospheres are sentinel beings continually working in concert with one another, measuring, responding, and adapting to the most subtle and dramatic of changes and the forces that drive them. They are imbued with a capacity to sense and act as an entangled and mutually beneficial heterogeneous set of systems. In concert, they act as a record of the histories of interaction, interference, reinforcement, and difference.

Geological Time

Covering 22% of the Australian continent, the Great Artesian Basin that feeds Kati Thanda is the largest artesian basin in the world. It is a register of ancient seismic shifts and sediment accumulation along with changes in the earth's magnetic field. Once the edge of the Gulf of Carpentaria, this sea body divided the driest continent on earth and is a remnant of an old oceanic plate. In its depths, it records the shifts in the ground from the past to the present through fluvial, aeolian, and lacustrine sediment deposits.

This vast landscape began forming more than 200 million years ago through major climatic oscillations that occurred during the planet's Quaternary Period, the third and final phase of the Cenozoic Era. The Quaternary Period, significant for glacial growth and retreat and for the extinction of large mammals and bird species, saw the spread of the human species across the planet. Today, this arid and semi-arid, unregulated, dryland river system supports threatened and endemic waterbirds and aquatic species. Its ecological systems respond to starkly divergent water flow regimes throughout the Great Artesian Basin, ranging from extreme to limited periods of flow, or even none at all. The cycles of boom and bust are punctuated by small and medium floods. Dulhunty's discoveries identify the lake as a "rain gauge" for the continent of Australia.[6] Through its palimpsest of geological traces, this landscape captures and discloses hundreds of thousands of years of climatic history in its sedimentary layers.

From 1972 to 1978, research geologist Dr John Dulhunty and his wife, Roma, returned each year to witness the transformation of this prodigious landscape. Together, they progressively documented the lake's geomorphology, including its phenomenal and rare transformation from a dry salt lakebed through an intense period of flooding to become the largest body of water on the continent, albeit a temporary one. They had observed the largest flood event in approximately 500 years.[7]

No water flows away from the surface of this endorheic lakebed. Water slowly flows into the basin predominantly from the northeast in Queensland and down through the channel country. As it languidly trickles across the ground, the water takes approximately 10 days to make its way across the lakebed. The groundwater flowing just beneath the surface of the lake continues to seep until it resides at the southernmost part of the lake, Kati Thanda North. Under this thick salt sheet lies an aquifer – a reservoir of briny water. As it emerges from under the crust, the water evaporates and its mineral matter recrystallises to thicken the salt sheet. There are places where the water escapes from the aquifer that lies about 1 kilometre under the Great Artesian

Basin. The water emerges through fissures as springs that are scattered throughout this landscape. The overflow of brine adds to the mineral salts and continues the ongoing transformation and formation of the salt surface.

When the optimal climatic conditions converge at the continental scale, the Kati Thanda landscape transforms, for a brief moment, from a dry salt bed into the country's largest inland freshwater body. In January 1974, a dramatic shift in weather patterns culminated in cyclonic wet monsoonal rains to the south and the highest-ever recorded monthly rainfalls in the north. Consequently, the northern rivers flooded and breached their banks, becoming kilometres in width as a continuous sheet of water. Before the water became too salty and when the weather conditions permitted, the lake glistened and shone like "fresh-cut topaz," transforming this desolate landscape from a salt sheet to a briny water surface.[8] The floodwaters of 1974 covered the lake and dissolved the Bonython Crust, just as the floods of 1950 dissolved the Madigan Crust. This was the birth of the future salt crust, a thicker, more scintillating surface emerging as a trace, a recorder of time, a manifold of the transformation of the geological strata. This manifestation occurs over time, and the process ensures that the northern lakes will shrink, and the aridity of the surface will result in a salt layer of varying thickness, so that a new salt sheet with a new name will arise.[9]

Microbial Life

A riot of sensation ensues upon the arrival of floodwaters from a basin that covers nearly 20% of the vast Australian continent.[10] The influx of water triggers a multitude of microbial transformations and the spawning and hatching of fish alongside the arrival of a vast number of bird species. Desert flowers spontaneously awake from dormancy, and an explosion of colour flows across the lake's usual vibrant gradient of pinks and blues. From the flow of water to the microscopic kaleidoscope of colours, this geological microbial bank can entomb a multitude of tenacious microorganisms into its thick surface for millions of years. The microorganisms calcify into salt crystals that continually resurface into the micro-biosphere.

The Archaea are single-cell microorganisms that are colloquially understood as demarcating the "limits of life on earth." They thrive in extreme environments, and along with algae such as Dunaliella salina and Nodularia spumigena, they infuse the sublime landscape with swirling colour gradients of pinks and greens. In contrast, Halobacteria prefer to have their molecules readily available for consumption. They scavenge the saline environment for carbon compounds that they oxidise.

A new world then comes into being, one inhabited by salt and soil dwellers. These microbes require us to see the world through different tools, time frames, scales, and life cycles, as they bear witness to the flurry of activity. To understand this world, we must simultaneously look through the lens of the microcosmos of the microbe and that of the larger stream of water and sedimentary flow. Alternative devices and acts of seeing are necessary, as "all our useful devices, our machines, only implement our acts."[11] Von Uexkull's concept of *"umwelt"* frames the subjective experience of an individual organism's environment. The tools of sense and structure equip organisms with the faculties to perceive and interpret while utilising their structural capacity in making their "close-knit worlds." The essential activity of these micro-beings is to perceive and act.

Historically, microbial communities have largely been overlooked due to a combination of geological, ecological, and evolutionary factors, rendering their members largely invisible. This invisibility makes it difficult to consider them, or separate them, from the visible matter of salt, soil, and strata. This raises broader questions regarding why certain entities become significant while others do not, and what it means for those beings that go unnoticed and therefore unknown. It also prompts ethical considerations that determine who or what "appears" to matter in ways that demand our attention. Consequently, we must examine what we can discern about beings that go unnoticed, or briefly pass us by, those that are invisible to the naked eye, and even those whose existence we are unaware of, such as unremarked species.

Kati Thanda attests to the power and complexity of the microbial world, the tiniest of organisms, and their immense capacity to shape the artesian basin and other life forms around them. This micro-world of incredible diversity contains a manifold of organisms that have evolved to survive in this harshest of environments. Despite their microscopic size, these microbes provide the foundation for life itself. They are a pertinent reminder that life is not just about what we can see with our naked eyes but also the invisible forces and flows that shape this ever-changing landscape.

The desert of Kati Thanda is devoid of figures and objects. However, its vastness and emptiness are not to be misconstrued as lacking liveliness or structure. This highly weathered ancient landscape is a space of radical indeterminacy and undecidability, where the traditional binary oppositions and hierarchies that structure our understanding of the world break down and reveal their inherent instability and contingency. In this sense, this landscape represents a space of deconstruction, where the fixed and stable meanings that we often take for granted are exposed as fluid and constantly shifting.

The predominant desolate nature that defines this landscape challenges conventional ways of understanding and relating to the world. It embodies a sense of immensity and openness that invites us to question our assumptions and explore the boundaries of our knowledge and understanding of how we experience the world. Moreover, it can be seen as a site of transformation and possibility, where new meanings and potentialities can emerge from the encounter with the unknown and the unfamiliar.

Mirage

A mirage is an endless and boundless experience. The shimmering afternoon sun beams down onto the landscape, which pulsates with the passing gales and shifting light.

A mirage alters the landscape. Chronological time as we know it has come to a standstill. Instead, a different life and rhythm are present, where increments, coordinates, and scale as purely quantitative measures cease to exist. Gaining an understanding of this landscape requires a new epistemology.

The hot desert air reflects and refracts the light, resulting in an optical illusion that resembles a body of water in the distant horizon. This complex assemblage of diverse material conditions and forces at play – the light, the air, and the landscape itself – makes Kati Thanda not only as a horizontal condition extending infinitely across the surface of the earth but also as a thickening condition and a four-dimensional one that incorporates space and time. It is a thickening of the ground that extends deep into the atmosphere.

The flatness of this landscape amplifies the material effects. A mirage is not simply an illusion but a material event that produces effects in this world, shaping the way we understand and interact with the phenomena of this landscape.

In this lively thickened condition, the light illuminates the landscape as a material force that interacts with its topography, atmosphere, and temperature to create a mirage, a sign that delivers, relays, and absorbs visible and invisible material conditions. "Mind and Body are seen as two levels recapitulating the same image/expression event in different but parallel ways, ascending by degrees from the concrete to the incorporeal, holding to the same absent centre of a now spectral – and potentialized – encounter."[12] For a few moments, a body in this landscape is alien, like a foreigner who has the inability to engage, which forces an engagement with a different set of positioning and orientation systems.

Within our field of view, a conflict of territory and structure emerges. Like a camera with an autofocus feature, we attempt to adjust and construct a frame with no apparent single focal point. From the onset, a seamless condition exists between the horizon and ground, between foreground, middle ground, and background. Zoom in, zoom out, or attempt to focus; none of these work. Conventions of projection are not relevant!

The gaze is not relevant in this situation. The eye of the observer shifts in this state. It is not the observer who determines the territories but the landscape itself. It is where the landscape becomes the observer. It looks back at what we have and what we have seen and is where its limits, its tendencies, its actions, and its effects are discovered and where temporary territories are made. Its fragments and relationships are the territories.

In this situation, different senses and modes of engagement are required. Seeing is not achieved by the eye alone but in association with touch, the other senses, and the body itself. The body in this landscape is a receptor and transmitter. A new set of tools is needed to see in this world, where the body in its entirety is not an object but an extension of the landscape. We need to see with all our senses, through the skin and with smell, sound, and taste. The body and the landscape are mutual extensions of each other.

Conclusion

Massumi suggests that our own "human" sensing of the world experienced through sensation involves a "backward referral in time."[13] Therefore, a sensation is organised recursively prior to being part of our conscious chain of actions and reactions. In this process, the smoothing over of an anomaly is made to fit our conscious requirements of continuity and linear causality.

This sensed landscape is a complex system of relationships. It has operational parameters, and their resultant physical expression is a modulator of form that is spatial, material, and temporal. It is intrinsically difficult to fully comprehend due to its dependencies, competitions, relationships, and other complex interactions between its parts and between the given landscape system and its environment. It expresses distinct properties that arise from these relationships as characteristics and tendencies. It also holds the potential for nonlinearity, emergence, spontaneous order, adaptation, and feedback loops, among others, that inform continual change. The landscape is formed from all these things interacting with one another that describes its "structure." The structure of the formation describes the motor of its substance that shapes its potential future "development" through change.

The reality of perpetual formation made visible and measurable through the sensed landscape describes an evolutionary model and behaviour of an ecosystem as the smallest measurable unit able to recycle biologically important elements nested within a network of ecosystems.[14] The parameters by which these units might be organised, defined, and subsequently evaluated typically fall under the umbrella of ecosystem services, which purport to provide social benefits such as pollination, clean drinking water, and recreational opportunities.[15] However, it is clear that the relationships formed between the sensed landscape, the information network, and the landscape architect hold the potential to shift this conventional paradigm. This includes a description of the work of nature as hybrid labour, "a collective, distributed undertaking of humans and nonhumans acting to reproduce, regenerate, and renew a common world."[16] This brings nature into the political economy, shifting it from a resource to be managed and plundered to a commons with equal rights, which aim "to call a more-than-human political collective into being, and to propose a relationship to nonhuman nature grounded in interdependence and solidarity rather than unidirectional management, ownership, or stewardship."[17]

The act of sensing and making the landscape is not a neutral activity, and therefore the process of representing forms a specific understanding of ecosystems and their processes. "Actant is a term from semiotics covering both humans and nonhumans; an actor is any entity that modifies another entity in a trial; of actors it can only be said that they act; their competence is deduced from their performances; the action in turn is always recorded in the course of a trial and by an experimental protocol, elementary or not."[18]

The tools for sensing a landscape and the techniques by which we deploy them have their own constraints that translate and transform information. The representations we make are constructed from a set of instruments, codes, techniques, and a lineage of conventions. Consequently, the worlds they describe and project are derived only from those aspects of reality susceptible to those techniques. These acts of sensing and seeing a landscape can formulate a view of what already exists and set conditions for new worlds to emerge.

Notes

1. Roma Dulhunty, *When the Dead Heart Beats Lake Eyre Lives* (Kilmore: Lowden Publishing, 1979), 3.
2. J. W. Gregory, *The Dead Heart of Australia: A Journey Around Lake Eyre in the Summer of 1901-1902, With Some Accounts of the Lake Eyre Basin and the Flowing Wells of Central Australia* (London: Murray, 1906), 145–155.
3. Donna Haraway, *Staying with the Trouble: Making Kin in the Chthulucene* (Durham: Duke University Press, 2016), 40.
4. Brian Massumi, *Parables for the Virtual: Movement, Affect, Sensation* (Durham: Duke University Press, 2002), 29–31, 58.
5. Lynn Margulis, *Symbiotic Planet: A New Look at Evolution* (New York: Basic Books, 1998), 132–134, 140, 158.
6. Kylie Carman-Brown, "A Tale of Extremes," *The People & Environment Blog*, *National Museum of Australia*, April 7, 2015, https://pateblog.nma.gov.au/2015/04/07/a-tale-of-extremes/.
7. Carman-Brown.
8. Roma Dulhunty, *The Spell of Lake Eyre* (Kilmore: Lowden Publishing, 1975), 42.
9. Dulhunty (1975), 1–3.
10. Massumi, *Parables for the Virtual*, 13–14.
11. Jakob Von Uexküll, "A Stroll through the World of Animals and Men: A Picture Book of Invisible Worlds," *Semiotica* 89, no. 4 (1992): 319.
12. Massumi, *Parables for the Virtual*, 32.
13. Massumi, 28.
14. Margulis, *Symbiotic Planet*, 132–134, 148–149.
15. Erle Ellis, *Anthropocene: A Very Short Introduction* (New York: Oxford University Press, 2018), 199. Kindle Edition.
16. Alyssa Battistoni, "Bringing in the Work of Nature: From Natural Capital to Hybrid Labor," *Political Theory* 45, no. 1 (February 2017): 6.
17. Battistoni, 7.
18. Bruno Latour, *Politics of Nature: How to Bring Sciences into Democracy* (Cambridge: Harvard University Press, 2004), 237.

SECTION II
Co-Creation

9
ORIGIN AND EVOLUTION OF BIODIVERSITY

A Story of Life on Earth

Christian Sardet

The story of life begins about 4 billion years ago while the Earth was cooling after taking shape and expelling a moon. Meteorites bombarded the planet, bringing rocks and chunks of ice that contained molecules including building blocks of life made in the cosmos. It's even a possibility that seeds of life came from outer space. We may never know, but the question of origins motivates our search to find manifestations of life on other planets.

Building Blocks of Life

All life is made of cells and all cells are made of an organized, dynamic cocktail of molecules – sugars, proteins, nucleic acids, lipids and metabolites. All these molecules are combinations of just a few elements: carbon (C), hydrogen (H), nitrogen (N), oxygen (O), phosphorus (P) and sulfur (S), commonly known as the CHNOPS. Cells are composed of molecules that combine these six elements as well as 20 minerals such as calcium, iron, iodine, magnesium, etc. In all, life relies on about two dozen elements out of a total of under 100 elements born from thermonuclear reactions and the transformations of hydrogen in the stars. Hydrogen, the first and most abundant element in the universe, appeared following the Big Bang 13.8 billion years ago. Humans and all organisms and cells are made of these elements. We are stardust.

The realization that all life is made of cells came slowly. Robert Hooke coined the term "cell" in 1664, because the tiny partitions he saw in a thin slice of cork through his microscope reminded him of the cubicles inhabited by monks or prisoners. At that time, observations through telescopes were upsetting the Earth-centered vision of the universe. The newly invented microscopes revealed that life was made of tiny entities called cells, and that fluids and liquids were filled with microscopic animal-like creatures: "animalcules". Two centuries of careful observations, dissections and staining of a multitude of plants and animals led to the formulation of a cell theory. The theory stated that all life was made of cells born from other cells by division. In the mid-1800s, Charles Darwin added the concept of evolution, giving a temporal dimension to the cell theory. Darwin had the intuition that all living beings came from a distant common ancestor, and that all organisms and species were related and branched from a shared tree of life. By the end of the 19th century, under the impetus of Ernst Haeckel,

DOI: 10.4324/9781003403494-12

representations of trees of life came to integrate unicellular organisms, single cells called "protists", at the base of the tree trunk. Bacteria and protists came to be considered as the ancestral cells whose evolution was responsible for the appearance of complex pluricellular creatures – animals, plants, algae and fungi. At the beginning of the 20th century, the emerging understanding of the chemistry of elements, atoms and molecules unveiled the physiological and molecular principles of respiration, fermentation and photosynthesis, revealing the importance of metabolism – the chemical transformations that drive all life and provide the needed energy.

The discovery of cells as the basic reproducing units of life did not immediately put an end to ancestral beliefs that life could arise spontaneously from inanimate matter. That idea was finally abandoned following the sterilization experiments conducted by Louis Pasteur in 1859 that killed bacteria. However, the belief that life can arise from non-life is still with us in a big way. It is at the center of our current ideas on the origins of life, which postulate a spontaneous genesis of cells from self-organizing and self-reproducing molecules and molecular structures. Most researchers think that this process of self-assembly of molecules into cells probably took place on Earth about 4 billion years ago. Still, some older primitive or prototypic form of life may have arrived from space in icy meteorites.

A Brief History of Life

There is general agreement that prototypes of cells – protocells – were present on Earth a few hundred million years after its formation. Among these protocells, an ancestral cell – nicknamed LUCA for Last Universal Common Ancestor – generated all life as we know it, life made of cells with the same set of molecules, and the same functioning and genetic principles, like sharing the same genetic code (DNA is transcribed in RNA, which is translated into proteins).

In the first million years of its existence, LUCA evolved into different types of single cells. Some survived and proliferated in the form of bacteria and another type of microorganism named archaea. An archaea basically looks the same as a bacteria but clearly differs from it in specific molecules and metabolic capabilities.

Bacteria and archaea are known as the prokaryotes – the cells without nucleus. They evolved together for more than 3 billion years up to this day. They colonize all environments from deep under the Earth up to the stratosphere and interact and exchange with all other forms of life in ecosystems. Two billion years ago, and maybe before, some species of bacteria and archaea cooperated and merged together to generate more complex cells.

The more complex cells that resulted from the merging of archaea and bacteria are known as eukaryotes. They are generally larger cells with well-defined interior compartments specialized in different functions like producing energy or packaging and exporting molecules. At first, eukaryotes were single cells, living as individuals or in association of same cells grouped into colonies. The eukaryote ancestor cells evolved into a variety of complex unicellar organisms known as protists whose typical representatives are yeasts, amoebas, diatoms and parasitic organisms such as the plasmodiums that transmit malaria. More than 800 million years ago, several types of protists transitioned from the unicellular to the multicellular state. This led to pluricellular organisms characterized by specialized groups of cells permanently attached to each other. These cellular communities evolved to become the ancestors of fungi, plants and animals. In the last 500 million years these macroscopic creatures have evolved and, with the help of microbes, they have colonized and transformed all the biotopes on the planet.

Scenarios for the Origin of Life

How life started on Earth remains a mystery. How could an inanimate world of molecules, minerals and ions self-organize into the first living cells? To answer this question, astrophysicists, geologists, chemists and biologists confront their visions and knowledge of planetary geochemistry, energy sources, molecular interactions and evolution. Some researchers believe that the first protocells were most likely born within the alkaline hydrothermal vents at the bottom of the ocean. Others argue that nuclear geysers or surface hot springs, or even evaporating puddles, were equally propitious environments. All these environments would have provided powerful and permanent sources of energy. They would also have favored the accomplishment of innumerable out-of-equilibrium chemical and energetic reactions, necessary for the emergence and proliferation of life. Among the plausible scenarios, let's imagine the appearance of life at the bottom of the ocean.

How Life Appeared in the Abyss

Let's go back 4 billion years. The planet is cooling. At the bottom of the ocean, volcanoes are spewing gases – hydrogen, methane, nitrogen, ammonia and sulfides, which percolate through the meanders of rocky microchannels. In contact with seawater charged with CO_2, huge mineral chimneys rise from the abyss. These rocky formations, riddled with billions and billions of micro-chambers lined with crystalline surfaces and metal complexes – iron, sulfides and nickel – are particularly conducive to the development of countless electrochemical reactions. The main idea behind this scenario is that a primitive vital energy force due to different concentrations of protons (H+) across semi-porous mineral walls favored the emergence of life. It presumably supported the reduction of CO_2 by hydrogen (H_2) and the production of a variety of carbonaceous molecules from simpler ones – methane (CH_4), formaldehyde (CH_2O), cyanides (HCN) and sulfides (H_2S). More complex molecules, a host of amino acids, sugars, lipids, nucleotides and other molecules of life were produced adding to the many organic molecules that arrived on Earth in meteorites from space.

Over millions of years, all possible chemical chain reactions took place within the innumerable micro-chambers of the abyss. That established a world of chemical chain reactions capable of continuously generating molecules of life: "a world of metabolism". Short chains of amino acids called peptides and nucleic acids such as RNAs as well as lipids that spontaneously assembled into micelles and membranes accumulated in the micro-chambers, testing all lasting self-sustaining combinations. The next step was the evolution of information-carrying molecules. Probably combinations of RNA molecules that associated with peptides. These RNA-peptide complexes acquired the capability of copying themselves with modifications and linking together amino acids into short proteins. This represented an early process of chemical evolution based on natural selection. A "world of RNA-peptides" endowed with the capacity to reproduce and to facilitate chemical reactions developed, becoming a source of information processing and organization. Then, a more stable cellular world based on DNA and a diversity of RNA and proteins endowed with enzymatic properties gradually established LUCA as the ancestor of all life.

From LUCA, the Ancestral Cell, to Bacteria and Archaea

The emergence of LUCA as the ancestor of all life implies that two worlds made of interacting molecules; the "world of metabolism" and the "world of RNA-peptides" came together and grew into a stable embryo of life. Perhaps these two worlds were first organized into molecular aggregates or condensates, acting as crucibles of chemical reactions and replications of information-carrying molecules. One can imagine that these life trials took place inside mineral micro-chambers and that the ancestral precursor cells were themselves dependent on electrochemical forces existing across the mineral walls. During millions and millions of years, all possible combinations between molecules and their associations were randomly tested until they could be sustained and perpetuated. Finally, time and time again, a "world of membranes" made of lipids associated with minerals and proteins encapsulated the molecular players of metabolism and information processing to generate the first protocells announcing LUCA. An embryo of life ended up inside functional membrane bubbles or vesicles that maintained and controlled the internal milieu of the cell. These encapsulations by membranes made up of a variety of lipid molecules gave rise to LUCA and the ancestors of archaea and bacteria. This is perhaps how bacteria and archaea acquired membranes made of the different kinds of lipids that characterize them.

In the abyss the ancestral cells did not need light to live. They drew their energy from inorganic molecules, minerals and gases abundant in their environment. Some bacteria and archaea continue to live today from the same gaseous and mineral resources at the bottom of the oceans. They even sustain rich ecosystems of tube worms and crustaceans in the abyss.

From Bacteria and Archaea to the More Complex Eukaryotic Cells – the Protists

During the first billion years of Earth's history, the ancestral bacteria and archaea evolved, diversified and proliferated on an oxygen-deprived planet. Then, about 2.5 billion years ago, some bacteria called cyanobacteria invented oxygenic photosynthesis, acquiring the capacity to live from sunlight. The cyanobacteria proliferated and produced so much oxygen that they changed the geochemistry of the planet – an event that has been called the "Great Oxygenation Event", abbreviated GOE. The soluble iron (Fe^{2+}) in the ocean oxidized and rusted. Oxygenation and the induced changes in CO_2 and methane concentrations caused a cataclysm – the total glaciation of the planet – 2.4 billion years after its formation. On this inhospitable "snowball Earth", the original microbial life was greatly reduced. It was a first massive extinction of unicellular life on Earth. Then, CO_2 spewed by the volcanoes for millions of years gradually warmed the planet. The surviving microorganisms adapted. The atmosphere and the climate, initially tropical, stabilized creating conditions conducive to the development of more complex cells resulting from processes of cooperation and domestication between different species of bacteria and archaea. We imagine that these microorganisms exchanged genetic material among themselves and through their viruses as their contemporary descendants continue to do today. Many species cooperated and complemented each other to survive. Local exchanges took place inside biofilms packed with different sorts of bacteria and archaea living together. In addition to the exchanges of molecules and genes, certain bacteria and archaea were able to merge, engulf each other and live inside each other until they domesticated each other. And on very rare occasions, these cellular chimeras gave rise to new kinds of viable cells. Over millions of years – and countless divisions, these cellular chimeras evolved into more complex cells with characteristic features of eukaryotes – larger cells with linear chromosomes inside a nucleus and a cytoplasm filled with numerous membrane-bound compartments called organelles. It is thought that the first membrane-bound organelles were mitochondria, which act as miniature power plants providing large amounts of energy to eukaryotic cells.

A popular explanation of the jump in cell complexity from procaryotes (bacteria and archaea) to eukaryotes is that a bacteria was domesticated by an archaea. The domesticated bacteria became a mitochondria and the presence of energy-producing organelles boosted the cellular complexity and size that characterizes eukaryotes. Another type of energy-producing organelle – the chloroplast – was later acquired from domesticated cyanobacteria, providing eukaryotes with the capacity to live from photosynthesis. Different types of protists acquired chloroplasts from other chimeric events. Since then, protists have diversified alongside archaea and bacteria while exchanging genes and other components with them and with viruses.

LECA, an Ancestor of All Eukaryotes

Exactly how and when did the first eukaryotic cell nicknamed LECA for "Last Eukaryotic Cell Ancestor" appeared is a source of debate. But, as with LUCA, the universal ancestor of all cells, a plausible account of LECA's genesis has emerged in the last decade. The story of the genesis of LECA involves two processes: that of syntrophy, which implies that two types of organisms feed one another, and that of endosymbiosis, a type of symbiosis in which an organism is ingested and then domesticated inside another. LECA, the ancestral eukaryotic cell, arose from bacteria and archaea that first associated (by syntrophy) and then, on rare occasions, merged and domesticated each other (by endosymbiosis) giving rise to a chimeric cell. It should be understood that this process took place over tens to hundreds of millions of years and that chimeric cells were only viable on very rare occasions. So much so, that it is considered that LECA is probably the sole ancestor cell of all protists, animals, plants, fungi and algae!

Several alternative scenarios are proposed. They involve two or three types of bacteria and archaea which carried out various exchanges and uses of gases, in particular hydrogen. One thing is certain: over time, a bacteria was integrated and domesticated inside a host cell – probably an archaea. The domesticated bacteria gradually became a mitochondria, an organelle which grows and divides in such a way that several mitochondria are transmitted to each daughter cell when the mother cell divides. For a majority of researchers, the presumed host cell that engulfed and domesticated a bacterium is an Asgard archaea. They reason that this particular branch of archaea, discovered in deep sediments, is closer to eukaryotes than bacteria. Asgard archaea are endowed with sets of genes and cellular structures characteristic of eukaryotic cells that do not exist in bacteria. It is therefore thought that between 1.5 and 2 billion years ago, an Asgard-type archaea engulfed and domesticated a bacterium and gradually became a mitochondria. But other scenarios challenge this view, and some researchers believe that the host cell could have been a large bacterium known to be able to engulf other bacteria. Not to mention that new species of bacteria are being constantly discovered including giant species with organelles.

What did LECA, the ancestral eukaryotic cell, look like? Born more than 1.5 billion years ago, LECA probably looked like an animal protist feeding on bacteria and archaea. This ancestral protist must have had mitochondria inherited from endosymbiosis with a bacterium. After many transfers of genes from the domesticated bacterium to the chromosomes of the eukaryotic cell, only a few genes of bacterial origin remained in mitochondria, which specialized in the supply of energy for the host cell. Then, hundreds of millions of years later, eukaryotic cells capable of capturing light energy appeared. These vegetal protists acquired in addition to mitochondria at least one chloroplast from an endosymbiosis with a cyanobacterium. This primary endosymbiosis event was subsequently followed by secondary and tertiary endosymbioses, during which protists domesticated other cyanobacteria or vegetal protists. These chimeric cells constituted new evolutionary plant lineages possessing chloroplasts and a variety of other organelles called plastids carrying pigments or nutrient reserves.

The origin and genesis of membrane organelles other than mitochondria and chloroplasts is more enigmatic. Opinions differ on the events that presided over the acquisition by the eukaryotic cell of the nucleus that surrounds the chromosomes with a double membrane. That double membrane – a selective barrier that controls what goes in and out of the cell nucleus – is part of a vast network of internal cellular membranes called the endoplasmic reticulum. One possibility is that giant viruses that cause the infected cells to build virus factories surrounded by membranes to produce and assemble new viruses, participated in the acquisition of the nucleus that personifies all eukaryotic cells.

From Protists to Pluricellular Organisms – Animals, Plants, Algaes and Fungi

The evolution and proliferation of life on Earth influenced the geochemistry of the Earth as well as the composition of the atmosphere and the climate. In turn, geochemical events and the diversification of natural environments shaped evolution and biodiversity. Due to the proliferation of oxygen-producing cyanobacteria, the first oxygenation event known as the GOE (Great Oxygenation Event) occurred more than 2 billion years ago. It introduced enough oxygen in the atmosphere to influence the evolution of novel and complex single cells – protists – living on respiration, fermentation and photosynthesis. The Great Oxygenation Event was followed, 1.7 billion years later, by another event called the "Neoprotozoic Oxygenation Event", abbreviated as NOE. It seems to have been initiated by the proliferation of photosynthetic protists and bacteria in the ocean about 750 million years ago. The NOE oxygenation caused two new glacial episodes which changed the existing biosphere. Atmospheric oxygen then increased to levels close to current levels while levels of CO_2 and methane fell, stabilizing the climate. An ozone layer was established, which protected from UV radiation the organisms living at the surface of oceans and on coastal areas and land where life was starting to spread. These conditions were favorable to the explosion of pluricellular life and macroscopic organisms. A diversified mixture of microorganisms, fungi, animals, algaes and plants spread on Earth – an early manifestation of the biodiversity existing today.

The millions of species of animals, fungi, plants, brown and red algae are complex beings made of many types of cells. Pluricellular organisms constitute three quarters of the planetary biomass, and for the last 600 million years they have shaped all marine and terrestrial ecosystems by developing complex relationships between themselves and with microorganisms. Some of these microorganisms live in symbiosis inside and on the surface of protists and pluricellular organisms constituting their microbiotes.

Pluricellular organisms are made up of eukaryotic cells like their single-cell ancestors, the protists. But unlike protists, which live as solitary cells or form colonies of the same cells, eukaryotic cells of pluricellular beings differentiate from each other and associate into a plethora of tissues and organs. Some of the cells – gametes or stem cells – can reproduce the whole organism. The common property of pluricellular organisms is that their cells are linked to each other. They are either permanently linked by openings that result from incomplete splits between dividing cells, or are joined by more transient links due to adhesive molecules secreted on their surfaces by the cells. The cells of plants, algae and fungi are permanently linked together via openings between cells, through which many molecules and, depending on the circumstances, organelles and even viruses circulate. From this point of view, animals differ from plants, algae and fungi, because the passages between animal cells are more restricted, except during the genesis of their reproductive cells – oocytes, spermatozoa and stem cells. Pluricellularity is therefore quite diverse, probably because it has arisen several times independently in different types of eukaryotic organisms.

The transition from the unicellular state typical of most protists to the pluricellular state characteristic of animals, plants, algae and fungi took place over hundreds of millions of years. The first pluricellular organisms were possibly fungi whose fossil record dates back more than 700 million years ago. The analysis of gene evolution suggests that pluricellularity occurred only once in animals, once in plants and more than a dozen times in fungi, brown and red algae. The trend toward complexity and gigantism is linked to the evolution of a multitude of specialized cells that may have represented advantages in terms of increased predatory, mobility and reproductive capacities. But it also seems to have generated problems of conflict between cells, as well as escapes from common rules controlling divisions and proliferation. Cancer cells are a manifestation of this.

Evolutions and Extinctions of Animals, Plants, Algaes and Fungi

Four billion years have passed since LUCA, the ancestral cell, generated different types of microorganisms made of single cells which, in turn, generated cellular communities that became pluricellular creatures like animals and plants. In terms of the evolution of biomass – the weights of the different forms of life – plants (450 Gigatons of Carbon) largely dominate fungi (12 GT C) and animals (2 GT C), while microorganisms collectively weigh about 82 GT C (bacteria, 70, archaea 7 and protists 5). The impact of humans has been such that, in terms of biomass of the mammals, the combined weight of wild mammals represent only 6%, livestock and domesticated species more than 60% and humans themselves 34%. In the aquatic environment of the oceans, planktonic microorganisms dominate in terms of biomass. Protists collectively weigh 2 GT C, and bacteria and archaea 1.8 GT C, while animals, plants, algae and fungi together represent only 2.8 GT C.

The evolution of the bigger life forms has not been a smooth linear process but rather a zigzaging path due to five periods of massive extinctions that caused some organisms to totally disappear from the tree of life. Living creatures were subjected to countless cataclysms caused by meteorite impacts, volcanic eruptions, fires and tidal waves, as well as climate changes and modifications of the atmosphere. The realization that two great oxygenation events – the GOE and NOE – 2000 and 750 million years ago had major impacts on life is recent. In contrast, the impact of extinctions has been known since the beginnings of paleontology in the 19th century. At the beginning of that century, Georges Cuvier perceived through the study of fossils that some animal and plant species had disappeared, and that others had appeared during catastrophic events that he called "revolutions of the surface of the globe".

Planetary catastrophes due to climatic upheavals, collisions with meteorites and/or volcanic eruptions have been the major causes of what we call mass extinctions. They are defined as such when they wipe out more than 75% of animals and plants in a short period of time. However, regional and localized extinctions are common and affect a variety of susceptible species all the time.

Since the Cambrian (541 to 485 million years ago), five major mass extinctions are recognized (−65/−201/−252/−365/−436 Ma). These extinctions sorted out which organisms survived, and each time they changed the course of evolution of the phyla displayed in the tree of life. For example, ancient arthropods called trilobites, the dominant marine animals of the Paleozoic era (−542 to −252 Ma), disappeared completely during the Permian extinction (−252 Ma). It was the most severe of all extinctions during which 90% of living organisms disappeared. The trilobites did not survive, but other arthropods – ancestors of crustaceans, arachnids and insects – were able to evolve, diversify and occupy most ecological niches. They have come to represent about 80% of the known animal species. In a similar way, the non-avian dinosaurs, descendants of the gigantic predators of the Jurassic (201 to 145 Ma), disappeared during the most recent mass extinction (Cretaceous extinction −65 Ma). However, some avian dinosaurs survived this extinction and became the ancestors of birds.

A sixth global extinction – the Holocene extinction – mostly due to human activities is progressing rapidly. One in eight bird species, one in four mammals, and 60% of all plants are endangered and threatened with total extinction in less than a century. The appearance of humans – a new species only 7 million years old – and the recent evolution and proliferation of human populations are having a huge impact on biodiversity and are shaping the future of life on Earth (Figure 9.1).

FIGURE 9.1 Origin and evolution of cell

FIGURE 9.1   Origin and Evolution of Cells

LUCA is thought to have appeared from the merging of the worlds of metabolism, of RNA-peptides and of membranes made of molecules – DNA, RNA, protein and lipid working together. LUCA evolved into BACTERIA and ARCHAEA, the PROKARYOTES – cells without a NUCLEUS having a single circular CHROMOSOME.

More than 1 billion years ago, BACTERIA and ARCHAEA merged into a more complex type of cell: a EUKARYOTE cell.

The first EUKARYOTES were single cells called PROTISTS. They all have several CHROMOSOMES within a NUCLEUS and other organelles such as MITOCHONDRIA and CHLOROPLASTS, which were inherited from ancient BACTERIA, engulfed and domesticated by another PROKARYOTE.

ANIMALS, PLANTS, ALGAES and FUNGI are all EUKARYOTES. They are the descendants of PROTISTS that evolved into pluricellular organisms.

10
ADRIAN PARR ZARETSKY IN CONVERSATION WITH JANET LAURENCE

Adrian Parr Zaretsky and Janet Laurence

APZ: The natural baseline species extinction rate is one species per year for every 1 million species. Although extinctions are part of the natural evolutionary cycle for life on earth, what we are currently experiencing is significantly higher than the background rate and that of the previous five mass extinction events. Today, the rate of species extinction is 1,000 to 10,000 times higher than the baseline rate. Over 35,000 species are threatened with extinction and over the past 500 years more than 900 have gone extinct. Your work grapples with the devastation extinction presents not just humanity, but all life on earth. In your view, what are the particular advantages of contemporary art engaging with a crisis, such as extinction?

JL: I think it is important that art does engage with these crises. We can be glazed by all the data and information that is out there, but art finds imaginative fresh ways to communicate such important information. Art reaches a broad and varied audience; it speaks to the heart and soul. Unless people feel emotionally, they don't grasp the desire and urgency to act. We need images, stories, actions, that bring an audience into action.

APZ: One of the biggest challenges the human population faces in effectively slowing the rate of species extinction is climate change. Over 1 million species are at risk because of an increase in average global temperature. For example, the sixth United Nations International Panel on Climate Change (IPCC) report warns that currently the stage is set for the earth to warm at least 1.5 degrees Celsius above the pre-industrial average within the next 20 years. More worrying, the earth is headed toward a median warming of 3.2 degrees Celsius by 2100 if greenhouse gas emissions are not dramatically curbed by 2025. Despite 194 countries signing on to the Paris Agreement in 2015 and collectively committing to pursue a limit of 1.5 degrees Celsius of warming, the majority of countries are yet to follow through with legislation and policies that align with pledges made at COP21 and subsequent COP meetings. At a time when so many have lost hope, experience frustration, or simply no longer trust the political process to effectively respond to climate change, art presents a different approach to the problem: exercising humanity's collective and individual capacity to imagine a world that is different to what currently exists. In this way art has the potential to expand the emotional and ethical range of people. How does your work engage with, or intervene in, the climate crisis?

JL: In different ways.

I am part of a group that I founded of environmental activists called Dirt Witches, made up approximately 80 women, mainly writers, artists, philosophers, all environmentalists. We create performances, make posters, petitions, and organize peaceful protests – often as artworks and art actions. We designed a t-shirt *I VOTE for the Trees* with a very prominent sustainable fashion group. The Dirt Witches also created a permanent Banksia Garden in the middle of the city as a restoration garden.

I held a two-week Requiem for the Bushfires in collaboration with the Sydney Environment Institute and the University of Sydney. There were many varied performances, including music, poetry, lectures, artworks, plant talks, and fire talks with indigenous groups.

Also, I created another performative work, a ritualistic processional work in the Powerhouse Museum Called Spelling Seeds in response to deforestation and the need for regenerating forests. Following that, another performance within the landscape of Bundanon Trust called Spells for Weather.

My exhibitions and performances create installations in order to engage a public to draw them into a space where my experiential work expresses our interconnection with and the fragility of our nature.

My work is also represented in Art Speaks Out and many environmental webinars and conferences where I speak and show my work as art that speaks for the ecosystems and earth rights.

I use aesthetics in the hope of creating memorable and evocative artwork. I believe an expression of beauty is important; it is after all what we're losing, the beauty and wonder of nature. To understand our loss, it is vitally important for us to see the wondrous and incredible complex system into which we are all interconnected.

My desire is to communicate through my work the places experiencing climate catastrophes and undergoing direct action, such as deforestation and mining which create massive extinction loss through the destruction of habitat.

APZ: If art can provide an alternative response to the absence of politicians to effectively produce policies and actions in response to the climate emergency, what is the nature of that response? How does it produce an alternative? In particular I am thinking about the installation format many contemporary artists work in, yourself included. Some might maintain that the immersive experience often central to the art installation format struggles to produce an adversarial aesthetic. Do you think an immersive experience necessarily forgoes an aesthetics of resistance and interruption? Does the political potential of art reside in an adversarial aesthetics, or do you think the transformative power of art lies elsewhere?

JL: This is complex to answer because it all depends on the intention of the work and how the work exists. Artworks can create a range of affective and effective responses. This is where the artists' personal language becomes important because the communicative power of art lies beyond the message. And, that is why art is so important – its ability to communicate emotions.

Immersive art relies on an audience to engage with it, not to repel, to take it in emotionally as well as intellectually. I think this is still a powerful way to engage requiring aesthetic entry. However, equally valid, but a totally different approach, is the one-off statement, a protest, a call out, an action. And there, I think an adversarial aesthetic works. Art invents surprising ways to speak.

APZ: It is not just other-than-human species that are facing extinction, *homo sapiens* faces extinction; indeed, as a species it could be argued that we are committing collective suicide if we continue to use dirty energy. Put simply, human beings have to collectively commit to making a difference, to produce a future that is different to the present and past. Yet ironically, this utopian turn to make the future different to what currently exists must also be informed by history and memory, in order to be realistic and effective. You strike such an incredibly provocative and poetic amalgamation of different temporalities in your work, bringing the affective power of memory into relationship with imagination and science. For you, does temporality function as a strategy of representation?

JL: Certainly, temporality is an important element in my work, in terms of what is rapidly changing, what we are losing, what we are experiencing, and our confusion of not knowing.In fact, more recently it's enabled me to play with the idea of Spells as strategies for change, because when all rationality has failed there is an attraction to what is seemingly irrational or to the supernatural. The idea is mischievous and playful, and that's full of possibility.

APZ: One of the greatest political challenges humanity encounters as it tries to slow climate change, species extinction, and ecological collapse is misinformation. Political leaders the world over have become increasingly more comfortable with lying. Disentangling fact from fiction is an important step in understanding how to respond to crises. Do you see art playing a role in sense making, helping people comprehend what is going on around them and building the will to hold our political leaders to account?

How do you hope to make a difference with your work?

JL: *Love in the time of extinctions, therefore, calls forth another set of questions. Who are we, as a species? How do we fit into the Earth system? What ethics call to us? How to find our way into new stories to guide us, now that so much is changing? How to invigorate love and action in ways that are generous, knowledgeable, and life-affirming?* (Quote from page 2 of "Wild Dog Dreaming: Love and Extinction" by UVA Press, 2011, by Deborah Bird Rose).

Art of all forms awakens our wonderings, imagination, curiosity, and love. Art enables us to explore our consciousness, seen and unseen. It can light up details creating a matrix that transforms old thoughts into new ideas.

Art is about transformation; we can't evolve without it. We need art now in this catastrophic time, as we need a totally changed Paradigm. Art is alchemical, enabling transformation though action; it is able to reenchant, reinvent, and regenerate.

TS Elliot said human beings cannot bear too much reality. So art enables the other space to come into presence. Art combines all different fields including history, spirituality, mythology, science, philosophy, poetry, the personal and the public – it can create a synthesis of these to create stories, actions, and images that draw people into it.

It is also able to speak directly the truth and can reveal the invisible that scientists and politicians and CEOs aren't free to reveal. The artist is a free speaker who can speak truths we cannot bear to hear.

Art speaks to the heart and the soul. It speaks a language of emotion and empathy to enable us to care; it is deeply important in our lives. We know well that joy of giving, loving, and caring, but now we need these things on a vast scale – like planetary gardening. Art can explore new ways to share our planet now with the other species that make us whole.

11

BIRDSONG AND THE TRANSPECIES AESTHETIC

David Rothenberg

Why is it we humans are so obsessed with animals? From the beginning we have always known they are both like us, and unlike us. Children are always told that we too are animals, while at the same time being yelled at, "stop behaving like an animal!" Clearly a mixed message that continues our whole lives, continues over the centuries. We always want to see just how animal and unanimal we are.

We are all born, grow up, live, love, reproduce, and die. It's easy to make it sound plain and obvious. It's easy to spend a lifetime unable to write enough on any one of those mundane and incomprehensible experiences. The animal question is at the heart of what it means to be human. We build civilizations as if to prove that we are better than all the other animals, and evolve culture recognizing that we are the most troubled of all animals.

Why so? We are the only creatures who doubt our very purpose, our very way to live. We are the only species unsure about how to proceed. All other animals know perfectly how to live, or never stop to question the reason why they live the way they do. Human essence is to be stuck inside philosophy. We are never sure, we are constantly arguing, we endlessly transform our world into something else that we hope will be better for us, all the while destroying the planet's ability to support all of us, humans and any other animals that still remain after the damage we have wrought.

It is hard and easy to write such things that have been said over and over for decades. But I am an optimist, I only keep offering up such critiques because I want our species to do good, to earn this right of specialness. Only we can destroy the planet, or save it.

I wrote a book about this twelve years ago called *Survival of the Beautiful*, in which I tried to convince you that evolution is far more than survival of the fittest, where each species evolved qualities useful for its continuation, such that every feature of every animal and plant is perfectly optimized for its fight against efforts to undermine it, where in every trait you see something pragmatic and useful.

I disagreed, citing Darwin's letter to Asa Gray about his favorite counter-example. "The peacock's tail!" to which our hero threw up his hands in horror. "It…makes me sick!"[1] Why? Because there seemed to be no reason for it. The poor resplendent bird, condemned to drag around this huge punishment of beauty, these amazing feathers that could be fanned out in magnificence. To scare predators? To lure in a mate? Maybe…but what a difficult way to further that. And what a poor description! First and foremost, this ultimate display of the best of all possible feathers is, before any mundane purpose, beautiful!

That's what we humans feel when we see it. Do the peacocks care about beauty as well?

Sickened by the peacock as much as the Argus pheasant and the birds of paradise, Darwin brought forth his theory of sexual selection in *The Descent of Man*, which includes such lines as, "birds have a natural aesthetic sense. That is why they have evolved beautiful feathers, and beautiful songs." In further detail he explains it is the less brightly colored and quieter female birds who have more developed minds able to discern which sounds and colors and shapes are more or less beautiful than others. Through their choices, they control the passing down of the aesthetic from generation to generation. The beautiful endures, evolves, survives. Survival of the weird, the wild, the extreme, the crazy, the magnificent—excess, ornament, astonishment, and delight!

Generations after generations of biologists have de-emphasized the impractical side of sexual selection and tried to subsume it under natural selection. Calling the peacock's tail and the nightingale's song indicators of some kind of genetic quality, "honest signals" of general health and good genes. Yet still, the bird with the most beautiful song is first and foremost the bird with the most beautiful song, not a musician coding some anonymous quality inside the specific ability. These birds sing, they grow beautiful plumage, beauty is inside their very essence.

Science sometimes tries to get back to this, still with mixed results. Richard Prum has promoted the idea in biology that beauty itself is an important engine of evolution, a tool of transformation that must be respected for its own way to push the metamorphosis of one species into another over the millennia.[2] This change is going nowhere else but through the changing mores of fashion, whose direction cannot be predicted. Evolution goes nowhere in particular, as ideas of beauty change only when a new possibility emerges (random mutation? a whim? an accident? a mistake? a sudden choice?) and it turns out the animals like it.

A few years ago a humpback whale in the Indian Ocean got a little lost, and rounded the straits between New Guinea and Australia, entering the Pacific. There he sang his Indian song for the Pacific whales, and they had previously heard nothing like it. Within a few years they were all singing this new song, an example where beauty in nature quite rapidly evolved. Was it only the shock of the new that impressed them? Or is there an essential sense in humpback whale song that one series of melodies is intrinsically better than another?

We don't know any animal's aesthetic sense that well, even our own. Who can say if Beethoven is better than Mozart? Most listeners would agree that that is the wrong question, that it is better to take each on its own terms and discuss what is good about each of them. Listeners of a more statistical bent would point out how many people listen to music many centuries after the works were composed, citing a certain amount of longevity as a sign that the work is good. Others might not be impressed by numbers, and believe instead that we shouldn't take them all that seriously. Most great works from throughout history have been completely forgotten.

In the animal world, to us at least, things seem clearer. Many of the bird songs we hear have been around for millions of years longer than the entire time humanity has been on the planet. So there must be *something* right about them!

We are the ones who worry, who change, who upend this restless Earth. Wait, though, the whales are doing it too. Even the white-throated sparrows. In recent years the familiar "Old Sam, Peabody," easy-to-remember mnemonic for that bird's characteristic song, is getting compressed, a new song coming down from Canada. These days it's more like "Old Sam Peabo."

The last syllable in the triplet is going away. Why? That is even harder to learn than why this species' song is so different from that of all the birds it is related to. Following Darwin, this is primarily an aesthetic question. If sexual selection is involved, that means it's changing because the females are starting to prefer the shorter, somewhat simpler song.[3]

But that might not be it. It's possible that aesthetic change just happens. The world is always changing, and if we listen close enough we can keep track of the change in our lifetimes, not just over the lifetime of millennia. Aesthetic attention isn't just about deciding what is good and what is bad, but before that it is simply paying attention. Noticing what is there, trying to take it in.

Humans do sometimes attend to the beauties of animal activities, bird songs in particular. In 2021 *The New York Times* reported a story on singing bird competitions in Surinam.[4] (Every few years this story seems to come back into the news.) Certain species of songbirds are prized for their melodies, and are set up in competitions against each other. Presumably this is a bit like what happens between males flaunting their talents in the wild, but there is something particularly interesting in this process when people get involved—*we* have to learn what is good and bad in a bird's aesthetic judgment. Are we learning from them, or deciding for ourselves?

Although I have not found exact criteria for how to judge the quality of a performance by a *picolet* (chestnut-bellied seed finch) or *twa-twa* (large-billed seed finch), I did manage to find the rules for how to judge the quality of the song of an American Singer canary, the breed believed to be the best caged singers in our own country:[5]

An American Singer should be neither too loud nor too soft. This is quite obviously very subjective (and highly sensitive to variables such as the acoustics of the judging room, the relative volume of other birds in a class, etc.). In broad terms, a bird which overpowers all the other birds in the room is too loud and one which cannot make himself heard is too soft.

Tone is defined as music or sound with reference to its pitch, quality, and strength. To those who prefer a simple explanation—such as myself—this refers to the bird's ability to sing on key with a beautiful, strong, rich fullness. A bird without good tone can sing the best song ever produced by a canary, but it just doesn't *sound good*.

Range refers to the lowest and highest pitches a bird can sing. Rollers sing in the low range while border canaries tend to sing in the high range. An American Singer should be able to sing both low and high notes.

Variety in simplest terms refers to the collection of distinct notes, tours, or song passages the bird sings. A bird that repeats the same limited number of notes and passages over and over again lacks variety. The term variety could also be used to more broadly describe the way in which a bird mixes the notes and tours- singing notes one way and then another and changing the order of passages and tours.

Melodiousness refers to the pleasing, harmonic way the bird puts his song together. The song should flow from one passage to another in a pleasant, coherent stream of sound rather than bounce from one sound to the next with little connection.

Showmanship is a vital part of the American Singer's performance. Often the major difference between a Grand Champion and a good singer is that the Grand Champion puts on a show- he is proud of his song and wants to be heard. He will not stand on the bottom perch, hide behind a water or seed cup, stand on the floor of the show cage- he perches confidently on a top perch, looks the judge in the eye, and *sings*. The judge cannot help but notice a bird who is a good showman.

There are 100 points upon which a performing canary is judged, 70 points of which refer to these quantities of sonic excellence described above. It is all turned into a sport perhaps? Are cellists auditioning for an orchestral position treated the same way? The Cincinnati Musicians' Union had this to say about candidates for a particular viola position in 2015: "We are definitely looking for the whole package deal. If I have to grade in all these areas, musicality, style, technique, you give certain grades in your head, you have your own scale. Some jury members concentrate on intonation, others on rhythm. So each juror brings their own list of priorities."[6]

This procedure is much like that applied to the canaries. In both cases, the judgment of the most beautiful is in the ears of the beholders, who are not the same as the performers, whose judgments might be quite different from that of the performers.

We really don't know what the birds most prefer, unless we try to decode the rules inside the music they make. Trying to take the music for what it is, to figure out how it is assembled, assembled as this thing that has evolved to a kind of perfection over millions of years. Such a result certainly deserves our respect, if we can figure it out.

Humans have a hard enough time agreeing upon our own standards of beauty. In music, is excessive ornamentation, vibrato, sweetness the best there is? "Serious" musicians never thought much about soprano saxophonist Kenny G, the best-selling instrumentalist of all time, but we know his "saccharine" sound is what more people like. Does the majority rule? Musicians prefer the saxophone stylings of John Coltrane, full of inventiveness and relentlessness, or Wayne Shorter, a master of the angular and of surprise. But the larger public finds a lot of their music difficult and obscure. Is the world of bird song aesthetics any simpler than that?[7]

Plenty of people still say that great art is good box office. It will sell. People will eventually come to love it even if it takes decades or centuries for them to get it. The music that is played generation after generation must be great, right? We remember Bach, Beethoven, Mozart. And I reckon there will be bands re-creating Pink Floyd's *Dark Side of the Moon* a century from now. The rebelliousness of the rock band will become classical music over time.

The bird songs we know are also a kind of classical music. The oldest enduring music we know. Most of these songs are simple and stylized, or "stereotyped," as the scientists like to say. But some are inventive, delightful, and complex, and we still can't figure them out.

Is it only because we don't bother to listen? I worked on a project with field biologist David Gammon, computational birdsong neuroscientist Tina Roeske, and myself the musician/questioning philosopher, to figure out some of the specific musicality inside the song of the mockingbird, probably the greatest avian musician recognized in North America.

The mockingbird's song is among the most stylish of voices in North American birds. It clearly follows rules and has a developed shape, form, and aesthetics. Surprisingly, no humans have tried very hard to articulate exactly what the bird is doing. In the course of our investigation, we found very specific moves the mockingbird makes, and very clear analogies to many kinds of human music.

With too complex a song to be evaluated like that of a canary, and too demanding to keep in a cage, mockingbirds have not caught on as pets. And seriously, who would want to think the song of any caged bird is better than its wild counterpart? That is an old, speciesist idea. The mockingbirds are easy enough to hear out there in the wild, and we know their songs have not evolved for us. But if we evolve our listening enough to take it seriously, we become all the greater for an expanded sense of what the beautiful can be.

It doesn't matter whether you particularly like or dislike the mockingbird's song. The point is to take it seriously as music, and figure out what is happening inside it. It surprised the three of us that no one seemed to have done that before. Gammon, the biologist who spends a considerable amount of time out of doors in the field, listening to and recording the birds, is probably the greatest expert on what the birds are actually doing. He can identify which kinds of sounds of which other species are mimicked in the mockingbird's song, and which sounds are wholly original and unique to the species *Mimus polyglottos* itself. The mockingbird does not simply imitate, or mock, but composes its song on the basis of very specific rules. It repeats its imitations two or three or four times, then *morphs* precisely from one sound to the next, changing some aspects of the sound while keeping others fixed.

In our paper for the erstwhile journal, *Frontiers in Psychology*,[8] in a special issue on how humans and animals learn songs and signs, we focused on just one aspect of this amazingly complex bird song, the transition from one syllable to the next, the specifics of the morphing. Like a canary competition judge, we first looked at many criteria, maybe nine, maybe ten, maybe fourteen, but then narrowed it down to four to make the analysis more clear. We came up with this:

The mockingbird may take a syllable and change the timbre of that syllable, morph one. It could keep the timbre but change the pitch, that's morph two. It could squeeze a sound and make it faster, that's morph three. Or he could slow it down, stretch it out. That's morph four.

Timbre, pitch, squeeze, stretch: Four simple methods of change, the beginnings of a key to the working of mockingbird music, from within. We know it's far too simple, but in simplicity of analysis comes the possibility of figuring out the complex.

It was Tina's job to run the numbers. Take hundreds of mockingbird songs and find out how often their transitions run like this. Timbre as the most popular choice by far, followed by pitch. Stretch and squeeze are neck and neck, equally less common. If we were to add a fifth it could be Contrast—at the end of a long series of morphs they like to conclude with a flourish, a sound totally different from what came before. Then a break, then I start the process again. Same rules, different sounds.

The transitions in mockingbird songs! How could this possibly matter to anyone! If we recognize the real beauty in the activities of other creatures, we have more reason not only to be convinced, or believe, our connection to all evolved forms of life which include ourselves, but to instantly *feel* this connection.

Aesthetics is not a matter of education, or only attention. It is a matter of resonance. Beethoven would understand (Figure 11.1).

We're not actually showing anything very complicated here, simply that human music also uses some of these basic transition strategies to hold onto something steady while changing another aspect of a sound or phrase.

Darwin revolutionized the notion of beauty in evolution by explaining how and why animals might evolve an aesthetic sense. Because said sense has not been considered especially *useful* by his disciples in biology over the centuries, we don't hear so much about it today. But we can learn what it is like to be a mockingbird by figuring out the preferences the mockingbird has evolved, the choices evolution has made upon its song through generations of female assessment of what he is doing.

Is there anything particularly good about the mockingbird's song, or of any bird's song for that matter? These songs have evolved. As Richard Prum likes to say, "beauty happens."[9] That simple fact is profound. It claims that the beauty of nature does change over time, usually a time scale far beyond the scale at which human culture or aesthetics evolves. By finding beauty in what nature has produced, we tap into a time deeper than that of any human experience.

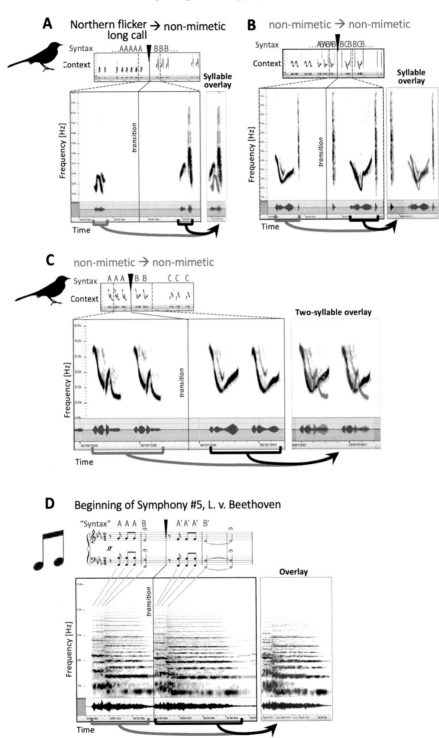

FIGURE 11.1 Morphing strategy "pitch"

So, that's why what some animals are doing can be considered music. How can we cross species lines to find a way to join in? That's what I spend most of my time doing.

The first example I usually give is this slowed-down, pitched-down recording of a single hermit thrush singing. This bird is famous among those listening for musicality in birds because its sound is unusually mellifluous, beautiful, and complex. Each phrase starts on a different note, and each phrase has an arpeggio-like quality, which means skipping adjacent notes in a scale, and there is a clear logic and placement of these tones but they are not quite those of a human scale. This means… the aesthetic of this species includes its very own scales, tuned its own special hermit-thrush way.

Plus this thrush leaves spaces between its phrases, unlike some other super-musical birds like the winter wren or the reed warbler. There is room to breathe, and to join in. Each phrase is a challenge because it comes with a whole other scale than anything we expect, but the number of intervals is manageable, closer to five to the octave, a pentatonic scale, than something like thirty-seven notes to the octave which would be very hard for a human to relate to.

What's interesting is that the intervals are not exactly what we expect. And there is clear musical inflection, so it can inspire a human musician to want to join in (Figure 11.2).

Every time I play with this recording, it sounds different. I want it to. It is a classic inspiration for improvisation, and my playing along with it builds on the thirty years I've practiced doing this. Somehow it never gets old.

Nevertheless, when I played it recently for a class of students at Pace University, I asked, "is this music?" and one immediately said, "No." How do we know if something is music? My usual answer is not the one I heard in school, which was "music is organized sound" or something like that. I prefer the idea that "it is music if someone tells you it is." You don't have to like that music, but the announcer is saying, "here is some sound, or something else—*consider* it as music, and see how you feel."

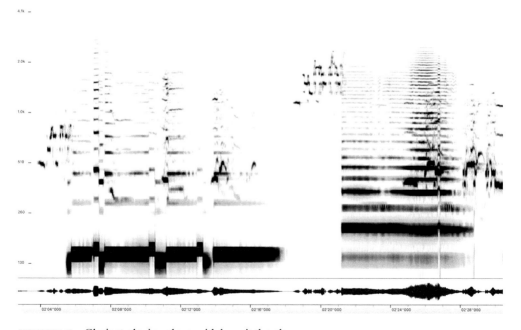

FIGURE 11.2   Clarinet playing along with hermit thrush

This student felt nothing. Probably because she heard sounds she could not place, nothing with tones she could align her expectations with: not major, not minor, no steady beat, no meaningful words. It inspired no emotion in her.

Maybe music needs to engender emotion, feeling, desire, delight. Not everyone is intrigued by the unexpected, the unknown.

You have to be a bit dissatisfied with the purely human world to want to make music with other beings, those you might never really understand.

The recording of the hermit thrush leaves space. Out in the actual world, before our manipulation, the songs of nightingales also leave space. Usually for other nightingales to join in, but we can also fill the spaces with our own sounds too.

Do nightingales *like* making music with people? The most rigorous study of nightingale response to playbacks of their own species' songs, conducted in Berlin in the 1970s by Henrike Hultsch and Dietmar Todt, discovered three ways a nightingale may respond to a strange new music in his midst. First, if he feels his territory is threatened, he will try to interrupt the unfamiliar sound—what the scientists called "jamming the signal"—thereby preventing any foreign message from coming through by getting in the way of it as much as possible. That's the aggressive response. But he may respond differently. A male nightingale who is confident in his territory, who doesn't consider you and your clarinet or iPad or voice or cello a threat, will listen to what you play, wait a moment, then respond with his own short song, and then pause again. If you give him some space, play a short phrase, and stop, the whole exchange is considered a friendly acknowledgment, with each musician trading ideas, leaving space for the next, accepting that we each have our place and our song.

Third, a nightingale who considers himself at the top of his game—the boss bird, the best singer in the whole park—will do whatever he wants, maybe interrupting, maybe leaving space, singing however long it pleases him, because you matter not in the least to him, convinced as he is of his greatness. He sings as if no one is there but himself.

We've all met musicians who fit into these three categories.

From a musical point of view, distinguishing between interruption and sharing could get quite blurry. What one person hears as jamming the signal could, to another, come across as just plain jamming, trying to make interesting music together. This is because music is far from a simple sign. It depends on what one believes music, in either a human or an avian context, to be all about. Perhaps artistry and form constitute not just an advertisement of territory and skill, but an attempt to work together to create something no one species could make on its own.

They may be challenging us, and deeming to impress. But they may also be enjoying themselves.

I never tire of that tale of the three ways nightingales respond to each other, it seems so achingly human. We humans are as prejudiced as any other species: when we find animals doing something like us, we think they're smart! Uniquely animal kinds of intelligence, like the social cohesion of ant societies, impress us a lot less. And of course, music isn't so much a matter of intelligence, but of feeling, emotion, aesthetic developments.

The nightingales certainly have that; they know some songs are better than others. We listen and listen, and still only dream of understanding what they know. As a transpecies performer, and someone fairly intuitive in my musical approach, the more I play live with these birds, the more I endeavor to fit in with them, as I felt in this one particular duet with a thrush nightingale in Helsinki, just before sunrise in June 2016, around 2:30am, when at the time I felt fed up with the whole project. Night after night trying to do this on two hours sleep, in the far enough north where it's hardly dark at all in the night. The bird doesn't want to be seen but we can easily see

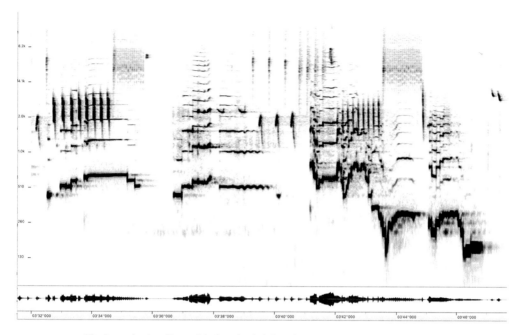

FIGURE 11.3 Clarinet playing live with thrush nightingale, zoomed in, Helsinki 2016

him, so instead of hiding in place on a perch he darts around, always on the move, impossible to film but easy to engage with because he does not stop, going on for hours, until the sun is ready to come up and too many sedge warblers, blackbirds, and robins make noise. He sounds frustrated, I felt frustrated, because I had been doing this for too many nights and had no idea what it all meant (Figure 11.3).

But then when I listened back I smiled—I was playing beyond what I knew and towards what the bird knows. I had heard my own music changing. And it keeps changing up until this day.

Playing with individual bird musicians is like playing duets. Playing with several birds is like chamber music, or perhaps since so much improvisation is involved it's closer to a jazz trio. With practice one learns the kinds of things the other musicians like to do, and we gain a sense of expectation, or anticipation. Still, even if one is prepared by experience, it's the possibility for surprise that makes the whole thing valuable. As Wayne Shorter once said, "you cannot rehearse for the unknown."[10] Still, we rehearse and practice all the time.

The next step is to figure out how to play with a whole ensemble of birds, and the best time to do this is the dawn chorus. Amazingly, science has not figured out why it is that birds sing so much at dawn. Is it too early for them to eat? Does sound travel better then? Are they announcing to the world they are present, here, ready to live? All these explanations have been proposed.

Modern humans may be less used to getting up just before dawn, when the birds are most active, but almost everyone who manages to do it feels a sense of rightness to the experience, that this is an essential time to be listening, present, and ready to join in.

Echoing the patience of the great Estonian nature sound recordist Fred Jüssi, who listened to nightingales for ten years before feeling ready to commit their songs to his Nagra tape deck, I urge you to spend at least a few mornings just listening so as to find your place. Any careful listener will soon hear what ornithologists are just beginning to admit—that birds at dawn don't

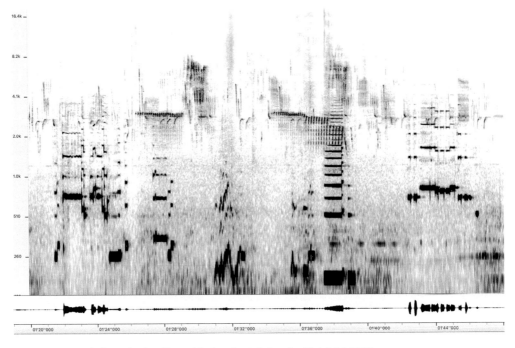

FIGURE 11.4 Clarinet playing live with thrush nightingale, Helsinki 2016

only listen to their own species, but to all the sounds in their midst, be they natural or human. They are inspired by all possibilities of sound. If you want to join in, I recommend hearing yourself as one voice in a vast improvising orchestra, finding your place by listening much more than you play, hearing your moment as one among a vast array of rhythms and tones as this excerpt clearly shows (Figure 11.4).

It is this metaphor of one hue among many that gives me the greatest hope for humanity better understanding our place in nature. Our lifestyles, our structures, our culture can fit into the surrounding world without destroying it, and this will be essential for us to find solutions to the looming climate crisis. Everyone knows this, but it can be so hard to actually know what to do.

I wouldn't say music or art can save the world, but it can put us into the mindset for the possibility of saving, by radically changing our whole worldview. Fitting into the natural world in a new way will require some adjustments, but it is absolutely something that our species has to try to do.

Music is not a language, not even a universal one. It is something beyond language, a form of communication that reaches from our world into the more-than-human world. It can connect us to nature, help humanity fit into the world surrounding us instead of beckoning us to destroy everything in our path. Music helps us find our place in the wider world.

"Birdsong," writes poet Joan Retallack, "entered our words and left with migratory echoes insufficiently dispersed. We are not designed to perceive most of what surrounds us or to fully understand the rest."[11] This may be why science can never be complete and why we also need the poetic to resonate with the beauty nature presents to us. But she knows better than to take that easy tack: "There's that conspicuous absence of real metaphor in nature. Sorry, I meant to say there's that conspicuous absence of real nature in metaphor."[12]

Either way, we lose. So why even try? Retallack explains. In addition to pushing science towards beauty, to its edge, she has to have the hubris to imagine we can *figure out* the beauty in nature, and at the same time, even in this great time of crisis for humanity and the planet, we see that beauty still can matter. It might even make sense to celebrate work that experiments, that pushes our sense of what in possible, in art as well as science.

She speaks of something called a *poethical wager*. The moral weight of believing in poetry?

[We are] a highly decorative, lightweight species that might seem almost like a biological whim, but of course, we know [we have] a very active place in nature. And that any individual, for reasons entirely unknown qua qua qua, could shift some ecological pattern—in a way noticeable or not to us, the "observant species." In other words, all one can do is take what is actually, in these terms, a very realistic, if improbable, chance that one's contribution might be useful. So that's it, the long and the short of it—my view of progressive action within a paradigm of chaos.… The *poethical wager* is just that we do our utmost to understand our contemporary position and then act on the chance that our work may be at least as effective as any other initial condition in the intertwining trajectories of pattern and chance. There's no certainty. One could, as John Cage said, make matters worse. But to make this wager is at least to step out into the weather of our times.[13]

So what are we to do with all these animal aesthetics? They are not preferences designed for us. We can invent our own preferences, and decide the song of the starling is unpleasant, or the wattle dangling off a turkey's neck to be red and unpleasant. That's one way to look at it. But we can also use all of our knowledge, and sensitivity, to imagine how they might appreciate it.

We can easily get closer to the animal world, by thinking not of the relativity of beauty, but the fact that it is situated within different worlds.

The individual mockingbird does not have completely free choice as to what sounds she finds most pleasurable in what the male sings. Her preferences have evolved because that's who she has developed to be. Humans may cause a lot of trouble, but we have one great aesthetic advantage: we can imagine what the world might look and sound like from another's point of view.

I started by announcing this is what makes humans exceptional: we can step back and consider what is beautiful from another species' point of view, even another person's! With that ability we could potentially offer the greatest amount of care for our world of any living species on Earth.

We have this choice.

Life on Earth is more than misguided purpose. We are most decidedly *not* just in it for ourselves. That's the other sense of human exception. We can step back from our own perspective, we may see what the world looks like from another species' point of view.

Take this risk and the world becomes all the richer the moment you do.

Notes

1 Letter from Charles Darwin to Asa Gray, April 3, 1860, Darwin Correspondence Project, no. 2743, http://www.darwinproject.ac.uk/entry-2743
2 Richard Prum, *The Evolution of Beauty* (New York: Doubleday, 2017).
3 Ken Otter, Alexandra Mckenna et al., "Continent-wide Shifts in Song Dialects of White-Throated Sparrows," *Current Biology* 30 (2020): 3231–3235.

4 Anatoly Kurmanaev, "A Battle of Singing Stars, with Wings and Feathers," *New York Times*, Jan. 14, 2021, section A, p. 10. https://www.nytimes.com/2021/01/14/world/americas/suriname-birds.html
5 http://www.americansingercanary.com/judging-of-the-american-singer-song.html
6 Janelle Gelfand, "Inside a Symphony Audition," *Cincinatti Enquirer*, Nov. 24, 2015. https://www.cincinnati.com/story/entertainment/music/2015/11/24/inside-symphony-audition/75478764/
7 See Penny Lane's wonderful film, *Listening to Kenny G* (HBO), and watch jazz critics squirm when forced to listen to the bestselling instrumentalist of all time.
8 Tina Roeske, David Rothenberg, and David Gammon, "Mockingbird Morphing Music: Structured Transitions in a Complex Bird Song," *Frontiers in Psychology* (2021). https://www.frontiersin.org/articles/10.3389/fpsyg.2021.630115/full
9 Richard Prum, quoted in David Rothenberg, *Survival of the Beautiful* (New York: Bloomsbury, 2011), p. 244.
10 Speaking in the film, *Language of the Unknown*. https://www.youtube.com/watch?v=sy17GpcZ79w
11 Joan Retallack, "The Ventriloquist's Dilemma," *Bosch'd* (Brooklyn: Litmus Press, 2020), p. 13.
12 Ibid.
13 Joan Retallack, *The Poethical Wager* (Berkeley: University of California Press, 2003), p. 45.

12

ENTANGLED INTELLIGENCES

Transpecies Dialogues of Art

Jiabao Li

Human-centric thinking is damaging the planet and other forms of life in ways that can't be repaired. Every 9 minutes, one species on this planet disappears forever. The challenge for humanity is to reimagine how we live through the lens of the environment and biodiversity; to shift our perspectives from a human-centered anthropocentric that is ego-centric and profit-driven industrialist to one that is a transpeciesist eco-centric worldview. This requires a fundamental change in perception that demands we rewire how humanity views nature and the anthropocentric hierarchy of intelligence that places humanity at the pinnacle. Through co-creation with non-humans, as an artist, I seek to tap into the wonder and needs of more-than-human species and share with the public through exhibitions and performances.

Squid Map

Working with Kewalo Marine Biology Lab in Hawaii with scientists who were studying the Hawaiian bobtail squid, I was struck by the monotony of the environments holding the squid. I began to wonder if I could make their environment more interesting by building a playground for them. I collected the white and black sands of the Hawaiian Islands, forming them into the shape of two countries, placing these forms inside the tanks where the squid lives. To my surprise the squid began carrying the sand from one side of the tank to the other, moving and reshaping it, the squid was, in effect, moving back and forth between countries, migrating across borders, burying itself under the sand, and using its little tentacles to push the sand onto its face as a camouflage.

Watching the squid reminded me of myself in many ways. I am an immigrant: moving between China and the US, carrying the baggage of two cultures, trying to assimilate to blend in much like the squid who used the sand to perfectly camouflage itself. Humanity has built borders between countries that often block or hinder animal migrations. Animals that would not naturally follow the delineation between countries have to recognize the difference and take a side.

After a month at the lab, the squid completely reinterpreted the maps and borders I had inserted into its environment. The map was now a squid-diagram formed from the squid's

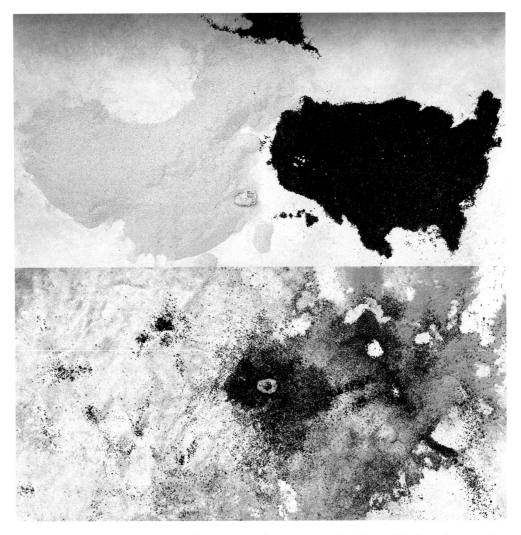

FIGURE 12.1 Squid Map. The upper image shows the map made of white and black sand prior to the introduction of the squid. The lower image, conversely, illustrates the same region after one month of squid habitation, showcasing notable alterations of the map and the borders.

perspective. The squid had in effect appropriated the human map and re-presented it on its own terms (Figure 12.1).

The upper image shows the map made of white and black sand prior to the introduction of the squid. The lower image, conversely, illustrates the same region after one month of squid habitation, showcasing notable alterations of the map and the borders.

While the scientists were experimenting on the Hawaiian Bobtail Squid, I witnessed the death of many squids—thousands of babies and some adults. One squid had died when a researcher was moving it to a new tank. The squid panicked and jumped from the balcony onto the second floor, landed on the ground, the panic it experienced prompting it to expel its ink. In response I created a squid funeral—a Gyotaku made from squid ink on rice paper. Japanese fishermen have long used Gyotaku to show off how big their catches are; however, in my practice I inverted the celebratory aspect of Gyotaku and used it as a platform to mourn the death of a marine life.

Chthulucene

Since the beginning of the Industrial Revolution, humans have become the primary drivers of environmental change, what is otherwise known as the Anthropocene. In my *Chthulucene* project I try to imagine a world where human beings are not the only important actors in the world by creating situations where diverse species are intertwined with one another. In this work we imagine a world where global warming causes sea levels to rise and all continents are submerged under water. As a highly intelligent species in the ocean, the octopuses have unlocked the optic gland gene that releases self-destructing hormones after giving birth. Now they can pass on the wisdom of their ancestors from generation to generation. The "Anthropocene" ends, and the earth enters the "Chthulucene." This work began with the prompt: What if the world is no longer centered around humans but octopuses?

When humans are on the edge of extinction due to environmental breakdown, how can we design an elegant extinction? What can we create so that the upcoming species, in this case the octopus, will remember us with tenderness and empathy? *Chthulucene* is a performance piece where dancers engage in an interspecies metamorphosis, attempting to connect with octopuses by learning their movements. The performance explores potential shifts from a worldview structured around human exceptionalism to one where humans must adapt and learn to embody a perspective beyond our species. This transition involves engaging with a distributed intelligence that necessitates moving away from the visual dominance typically associated with human experience and cognition, towards a more tactile perception (Figure 12.2).

A still image captured from a video showcases two dancers engaging in contact improvisation. This unique dance form serves as a metaphorical exploration of adapting and learning to

FIGURE 12.2 Chthulucene. A still image captured from a video showcases two dancers engaging in contact improvisation. This unique dance form serves as a metaphorical exploration of adapting and learning to embody a perspective beyond our species, particularly embodying characteristics of an octopus, thus pushing the boundaries of human-centric perceptions.

embody a perspective beyond our species, particularly embodying characteristics of an octopus, thus pushing the boundaries of human-centric perceptions.

The piece considers what this world would be like from the experiences and perception of an octopus. Maybe this means there would be bio-engineered humans with enlarged organs specialized in their functions to adapt and survive under water. Or it might involve octopus archeologists scanning through the remains of the Anthropocene – trash from the ocean.

Squeeker: The Mouse Coach

In the realm of scientific research, mice have been utilized as crucial models to understand behavioral change and habit formation in humans.[1] Scholars have identified vital neural circuits and mechanisms via numerous studies, informing us about behavioral change and new habit formation processes. For instance, it has been indicated that the reward system of the brain is pivotal in habit formation, and repetitive behaviors could lead to modifications in neural circuitry, making habits more ingrained.[2] Moreover, mice have also been employed to investigate environmental influences, such as stress and social interaction, on behavior change and habit formation. Every year hundreds of millions of mice get fat and stressed, or fit and healthy, all in the name of helping us humans to comprehend ourselves better. What if we switch the dynamic and let the mouse control our health?

Observing my two pet mice in our shared living space, I noticed their weight gain when their spinning wheel malfunctioned, an event that echoed my personal struggle with the absence of regular exercise. To tackle my dwindling motivation, I drew inspiration from my pet mice, transforming them into personal fitness coaches. I engineered a spinning wheel integrated with sensors and an App that sent me a notification to run each time my mouse coach commenced their workout. I devised a rewarding system where if I matched the distance they covered, they would receive a treat. And I was allowed to scroll on social media, with the scrolling distance the same as my running distance. This approach leverages the theory of the habit loop, whereby Cue, Action, and Reward are vital for the establishment of healthy habits.[3,4] In this scenario, the mice's running serves as the cue. The reward is the mouse coach receives a treat and I received the dopamine rush from scrolling on social media. Considering that mice run around 4–20 km per day, maintaining their pace essentially resembles half-marathon training. To testify the mouse coach's training on me, I successfully ran the Austin half-marathon on Feb. 18, 2024 (Figure 12.3).

A still image captured from a video highlights a mouse running on an exercise wheel equipped with sensors to measure the distance traversed by the mouse. When the mouse runs, the sensors trigger a notification to the artist, prompting her to mirror the mouse's action and engage in her own physical activity.

Nocturnal Fugue

Our understanding of animal communication, specifically among bats, has dramatically improved due to advances in deep learning technology. Yossi Yovel's Bat Lab has made significant strides in this area by monitoring Egyptian fruit bats' vocalizations for over two months, employing a voice recognition program to analyze their sounds.[5] The algorithm implemented by the team has been instrumental in categorizing a majority of the bats' sounds, revealing a more complex language than previously recognized. The bats demonstrate the capability for

FIGURE 12.3   Squeeker: The Mouse Coach

individual identification, engaging in disputes over food, and vocal learning. Interestingly, mother bats decrease their pitch when communicating with their offspring, an intriguing contrast to the heightened pitch found in human baby talk.[5] Such discoveries have been facilitated by deep learning tools and technologies, which offer unprecedented opportunities for communication with bats and other animals. Beyond mere understanding of bat language, researchers are now attempting to use generative AI to respond to bats in ways previously thought impossible.

My personal engagement with bats at the Austin Bat Refuge has offered me profound pleasure and fulfillment. Rehabilitating injured bats and recording their captivating sounds invokes a sense of awe and fascination. When two million bats take to the night sky from the iconic Austin Congress bat bridge, their ephemeral symphony of social calls and echolocation generates a detailed map of their world. To further engage with these intriguing creatures, I have embarked on a collaborative journey with a sound artist, Matt McCorkle, to craft a musical symphony composed of bat sounds. What would a baby bat lullaby sound like? Would a song made of bat mating calls be a love song? In "Nocturnal Fugue," we created an immersive world of bats through spatial sound, immersive projections, and haptic vibrations. Their social vocalizations transform into evocative music set to imaginative digital recreations of their natural habitats. Each digital habitat crafts a different sense of scale and time, creating a mood centric to the social vocalizations that are being heard and felt. People can explore the identity of bats through their isolation, sleeping, kissing, mating, feeding, grooming, and fighting vocalizations.

Inspired by Thomas Nagel's philosophical inquiry in "What Is It Like to Be a Bat?", "Nocturnal Fugue" delves into the enigma of truly comprehending the experience of another species. While we may grasp the superficial metadata of bat behavior, the true content of their communication remains elusive. Can we ever fully understand the language of bats, even with the aid of artificial intelligence? We might not be able to find many words from the bat world in human's dictionary.

Once a Glacier

Glaciers grow, store memories in ice cores, and transform throughout their lifetimes. To study these phenomenal entities, I relocated to Alaska and collaborated with local glaciologists and community members, aiming to understand the dramatic retreat of glaciers observed over the past century. By translating obscure climate change graphs into more relatable narratives, I strive to challenge audiences to comprehend the intricate beauty and irreversible transformations of nature.

Taking data from the last six decades regarding glacier melting, I created a composition of music and dance with local musicians who have personally witnessed the Mendenhall glacier's regression.[6] We performed the composition, "Glacier's Lament," on the glacier itself, accompanied by the natural soundscape. In one expedition, I retrieved a fragment of glacial ice, an act symbolic of a futile attempt to preserve a piece of the rapidly vanishing glaciers. I endeavored to cultivate this ice piece within my freezer, nurturing it with water. As I listened closely, I discerned a plethora of sounds reminiscent of a forest within the ice. This inspired the creation of a virtual reality film, "Once a Glacier," narrating the story of a girl's relationship with a piece of glacial ice and her futile struggle to protect this remnant of what was once a vast glacier.

In the Inupiaq tradition, glaciers are revered as keepers of past memories, communicating these through their distinctive songs. The climate crisis, however, has threatened these sung histories with extinction, as the demise of glaciers unfolds before our eyes. A poignant example is a unique chirping sound, emitted when glacial meltwater bubbles burst into the air. This sound lacked an English equivalent, and I discovered that in the local indigenous language, Tagish, it was referred to as "the sound of the break-up of ice."[7]

Scientists have employed hydrophones to monitor the rate at which glaciers are melting by detecting the sound of these bursting bubbles.[8] By estimating the number of bubbles released into the water through their distinctive sound, they can infer the melting rate and subsequently, the speed of glacial retreat.

This process underscores the dual facets of "deep listening"—a time-honored and vulnerable indigenous tradition and a cutting-edge technique bolstered by artificial intelligence. Indigenous communities have long co-existed with nature, cultivating a profound attenuation to non-human sounds and interpreting them into meaningful stories and tales. Artificial intelligence enables us to decode nature's language, comprehend its current state, and form new, technologically-derived narratives.[9]

Transpecies Co-creation

The concept of transpecies co-creation—a collaborative artistry between humans and non-human species—offers a unique and profound way to appreciate life's diversity on Earth. This collaboration involves both species contributing to the creation of the artwork, where humans sometimes construct the scaffold upon which the animal designers build.

One example of this unique partnership is the "Bee vase," designed by Libertíny. This piece involves the designer creating a vase-shaped scaffold that bees fill with their hive layer by layer. This method results in a vessel that poetically embodies life's cycle, beginning with flowers that nourish bees, and culminating in a vase designed to contain flowers.[10] Tomás Saraceno's Spider web is another example of co-creation, where he places solitary spiders in transparent cubes and allows them to spin webs. Then, he introduces a colony of social spiders that build on top of the original spider's creation, forming new intricate structures. The works' titles credit the names and species of the spider collaborators who came together to tune their strings.[11] In the Silk Pavilion, created by Neri Oxman's group, the insects acted not only as construction workers but also as designers, working in tandem with a human-made structure that guided their movement and silk deposition, resulting in an enhanced form. In contrast to traditional silk harvesting methods that involve boiling the larvae alive in their cocoons, this process allows the silkworms to live and metamorphose in relative peace.[12,13]

Aki Inomata's co-creation with beavers raises questions about the agency and authorship of non-human species in collaborative artwork.[14] Beavers chew on trees to sharpen their teeth and construct dams, resulting in unique markings on the wood that Inomata sees as sculptures. She commissioned a carpenter to replicate the beaver's work on a human scale, which led to new interpretations due to the carpenter's hand carving. Inomata also used a robot arm to carve another copy, prompting the question of who the "artist" is: the beaver, the carpenter, the robot, Inomata, or even the wood itself? This work highlights the challenge of co-creating with non-human species while acknowledging power hierarchies. Inomata's examination of the human-pet relationship is represented in her work that involves collecting her dog's and her own hair to make coats for each other. This exchange of hair represents the embodiment of bonds, like a memento or a vow.

These collaborations raise pertinent questions regarding the agency and authorship of non-human species in the artistic process. The ethical considerations of treating animals as equals and the necessity of ensuring no harm to living systems become paramount. It is essential to appreciate the complexities and agency of living systems and acknowledge their unique intelligence. Additionally, the co-creation process should be driven by the process and not predetermined by one side. Consent, a concept normally associated with human interactions, becomes crucial when collaborating with living systems.

The Role of AI and Technology

The relationship between humans and non-human beings has long been a subject of philosophical inquiry, with views and attitudes evolving over time. As we look back through the annals of history, we can see that even ancient civilizations held beliefs about animals that shaped the way they were treated and perceived. Philosopher Descartes famously espoused the idea that animals were nothing more than complex machines, without consciousness or subjective experiences.[15] In other words, animals were not capable of feeling pain, pleasure, or any other emotions, and were not deserving of moral consideration. His ideas were based on an anthropocentric view of animals, where humans were seen as the most important and valuable beings in the world, and all other creatures were judged in relation to them. His view has been widely criticized by modern philosophers and scientists.[16] Evidence suggests that animals have complex cognitive, emotional, and social lives.[17] For example, research has shown that many animals, including primates, dolphins, and birds, are capable of using tools, communicating with each other, and exhibiting a wide range of emotions.[18] Elephants, for instance, have been observed mourning the deaths of their kin, while dogs have been shown to have a sense of fairness and can even experience guilt.[19,20] In addition, they have been shown to exhibit acts of selflessness and display remarkable social intelligence, even showing love for other species. We have come to challenge many of our preconceived notions about our supposed human exceptionalism. This new understanding has enriched our world and illuminated a more wondrous, interconnected, and awe-inspiring perspective on non-human species.

Technological advancements enable a deeper appreciation for nature. For instance, Stanford professor Manu Prakash's Foldscope, a low-cost microscope, fosters a global community of explorers.[21] Platforms like iNaturalist connect people to scientists, aiding the identification and understanding of surrounding flora and fauna, while also contributing to scientific research.[22]

In addition to tools that allow us to appreciate nature at different scales, there are exciting developments in the field of animal behavior and communication research. Through deep learning and innovative algorithms, we are beginning to interpret the behaviors and language of other species, which may have important implications for their conservation. For example, in the Panda Project, researchers used deep learning to identify pandas' activities based on their sounds.[23] Such understanding could lead to improved conservation efforts. The Earth Species Project aims to use AI to decode non-human communication, enabling us to monitor populations, recognize diversity, and understand human impact on various species. They believe AI is akin to modern optics, revealing patterns at an unprecedented scale, and offering a new lens through which we can see our planet and humanity's place within it.[24]

Humans have a tendency to optimize for one factor rather than the entire ecosystem, and this is where AI can be a valuable tool. The Pollinator Pathmaker project, developed by artist Alexandra Daisy Ginsberg and Google Arts and Culture Lab, demonstrates this beautifully.[25] By using a unique algorithmic tool, the project helps us design gardens not for humans, but for pollinators such as bees. Since pollinators see colors differently and emerge in different seasons, a garden designed for them may look quite different from one designed for us. FluentPet is a company that creates communication buttons for pets, which enable animals to express themselves by pressing the buttons.[26] These buttons have transformed the way people interact with their pets, allowing them to better understand and communicate with their furry friends. By giving pets a voice, these buttons open up a new world of possibilities for human-animal relationships.

Animal Influencers

The repeated prophecies of vanishing coastlines and species extinctions have led to an insidious phenomenon known as climate change fatigue. Amidst such apathy, the idea of Animal Influencers emerges as a beacon of hope. This concept was the focus of our panel "Animal Influencers as a Way Out of Climate Fatigue" at SXSW 2023. I invited bat conservationist Merlin Tuttle, OctoNation founder Warren Carlyle, and artist Virginia Lee Montgomery to this conversation.[27] The underpinning of the animal influencers approach is to inspire positive change and spark curiosity, rather than fueling the discourse with anger and anxiety.

Merlin Tuttle has long advocated for positive interactions between humans and the natural world as the key to successful conservation efforts.[28] Tuttle's philosophy is that building relationships with individuals and communities is more effective than combating those threatening bat habitats. His organization, Merlin Tuttle's Bat Conservation, dispels misconceptions about bats and promotes their ecological and economic importance, leading to the creation of bat-friendly communities and policy changes favoring wildlife conservation. Similarly, OctoNation uses octopuses as ambassadors to inspire awe and curiosity about the ocean. The fan club generates income for scientific research on octopuses and leverages merchandise, like stickers, to educate the public about different species of octopuses.[29] This approach aligns with Jacques Cousteau's belief that education fosters understanding, love, and protection of the environment.[30]

The influence of animal ambassadors extends far beyond the realms of animal conservation and ecology. Brands can leverage the power of animal branding, using representative animals in their logos, such as Twitter's bird, Evernote's elephant, TripAdvisor's owl, Puma, and Dove. For instance, Bacardi Rum's logo features an easily recognizable bat. Bacardi has been working with Merlin Tuttle to save bats and has a long-standing commitment to promoting bat conservation to the public.[31] As part of their efforts, Bacardi published one of the first educational brochures that introduced people to the truth about bats, dispelling misconceptions and highlighting the conservation needs of bats.

In a world where celebrity endorsements come and go, animals remain the ultimate timeless spokespeople, capturing hearts and inspiring awe for generations to come.

Tencent, a video game and internet industry leader, utilizes a penguin as their logo.[32] I collaborated with them to utilize the endearing penguin branding to ignite a curiosity for these beloved creatures and increase awareness of the dire threat of extinction facing the Emperor Penguin as a result of climate change. The consequences of altered sea ice in Antarctica have affected the quality and availability of breeding and feeding habitats for multiple penguin populations, including temperate species impacted by changes in sea surface temperature and global climate oscillation like the El Niño Southern Oscillation.[33] Unfortunately, climate change remains elusive and ambiguous to many. By endowing it with a charming representative, we can build an emotional connection with the issue and foster greater action towards protection. The animal world abounds with remarkable influencers who evoke wonder and admiration. Appreciating nature is the initial step towards preserving it.

There are numerous types of intelligence beyond human intelligence, such as the distributed intelligence of octopuses, the swarm intelligence of ants, bees, and birds, the collective intelligence of mycelium, and microbial intelligence. Co-creating with non-human species allows us to probe different intelligences, shift perspectives, and interrogate ourselves. If the entire world is a projection of our perception, how do we see the world from a cephalopod's perspective? What can we create from that viewpoint? How can we co-design with them and for them? To collaborate, we need to go beyond passive observation and deeply understand non-human intelligence, behavior, and agency. We need to respect, learn, connect, and share the umwelt. Recognizing these connections reveals how we are part of many continuums of intelligence.

Notes

1. Bilkei-Gorzo, A. "Genetic Mouse Models of Brain Ageing and Alzheimer's Disease." *Pharmacology & Therapeutics* 108, no. 3 (2005): 519–566.
2. Graybiel, A.M., and Grafton, S.T. "The Striatum: Where Skills and Habits Meet." *Cold Spring Harbor Perspectives in Biology* 7, no. 8 (2015): a021691.
3. Duhigg, C. *The Power of Habit: Why We Do What We Do in Life and Business* (New York: Random House, 2012).
4. Clear, J. *Atomic Habits: An Easy & Proven Way to Build Good Habits & Break Bad Ones* (New York: Avery, 2018).
5. Yovel, Y., Falk, B., Moss, C.F., and Ulanovsky, N. "Active Control of Acoustic Field-of-View in a Biosonar System." *PLoS Biology* 8, no. 9 (2020): e1000545.
6. Motyka, R. J., O'Neel, S., Connor, C. L., and Echelmeyer, K. A. "Twentieth-century Thinning of Mendenhall Glacier, Alaska, and Its Relationship to Climate, Lake Calving, and Glacier Run-off." *Arctic, Antarctic, and Alpine Research* 34, no.1 (2002): 83–88.
7. Arnold, S. "Future Rivers of the Anthropocene." *Open Rivers*, no. 15 (2020). Accessed June 2, 2023. https://openrivers.lib.umn.edu/article/future-rivers-of-the-anthropocene/.

8 NPR. "Scientists Are Using Microphones to Determine How Fast Glaciers Are Melting." Last modified October 17, 2022. Accessed June 2, 2023. https://www.npr.org/2022/10/17/1127158854/scientists-are-using-microphones-to-determine-how-fast-glaciers-are-melting#:~:text=Scientists%2520traditionally%2520have%2520relied%2520on,using%2520underwater%2520microphones%2520called%2520hydrophones.
9 Bakker, K. *The Sounds of Life: How Digital Technology is Bringing Us Closer to the Worlds of Animals and Plants* (Princeton: Princeton University Press, 2022).
10 Libertiny, T. "The Honeycomb Vase 'Yellow'." Tomas Libertiny. Accessed June 2, 2023. http://www.tomaslibertiny.com/the-honeycomb-vase-yellow.
11 Saraceno, T. "Hybrid Webs." Studio Tomás Saraceno. Accessed June 2, 2023. https://studiotomassaraceno.org/hybrid-webs/."
12 Oxman, N. *Material Ecology* (New York: The Museum of Modern Art, 2020).
13 Oxman, N. "Silk Pavilion II." Neri Oxman Projects. Accessed June 2, 2023. https://oxman.com/projects/silk-pavilion-ii.
14 Inomata, A. "Why Not Hand Over a "Shelter" to Hermit Crabs?" *Aki Inomata*. Accessed June 2, 2023. https://www.aki-inomata.com/works/hermit_2009/.
15 Mahaffy, J. P. "Descartes' 'Animated Machines'." *Nature* 5 (1871): 62–63. Accessed June 2, 2023. https://www.nature.com/articles/005062b0.
16 de Waal, F. Are We Smart Enough to Know How Smart Animals Are? (New York: W. W. Norton & Company, 2016).
17 Ahmed, A., and Claudia C. "Animal Sentience and Consciousness: A Review of Current Research." *Nuffield Council on Bioethics*. Accessed June 2, 2023. https://www.nuffieldbioethics.org/assets/images/Animal-sentience-and-consciousness-review.pdf.
18 Bekoff, M. "Animal Emotions: Exploring Passionate Natures: Current Interdisciplinary Research Provides Compelling Evidence that Many Animals Experience Such Emotions as Joy, Fear, Love, Despair, and Grief—We Are Not Alone." *BioScience* 50, no. 10 (2000): 861–870. Accessed June 2, 2023. https://doi.org/10.1641/0006-3568(2000)050[0861:AEEPN]2.0.CO;2.
19 Douglas-Hamilton, I., Bhalla, S., Wittemyer, G., and Vollrath, F. "Behavioural Reactions of Elephants towards a Dying and Deceased Matriarch." *Applied Animal Behaviour Science* 100, no. 1–2 (2006): 87–102.
20 Horowitz, A. "Do Dogs Feel Guilty?" *Scientific American* (blog). Last modified October 18, 2012. Accessed June 2, 2023. https://blogs.scientificamerican.com/thoughtful-animal/do-dogs-feel-guilty/."
21 Cybulski, J.S., Clements, J., and Prakash, M. "Foldscope: Origami-based Paper Microscope." *PLoS ONE* 9, no. 6 (2014): e98781.
22 "iNaturalist." Accessed June 2, 2023. https://www.inaturalist.org/.
23 Wang, Hanlin, Jinshan Zhong, Yingfan Xu, Gai Luo, Boyu Jiang, Qiang Hu, Yucheng Lin, and Jianghong Ran. "Automatically Detecting the Wild Giant Panda Using Deep Learning with Context and Species Distribution Model." *Ecological Informatics* 72 (December 2022): 101868. Accessed June 2, 2023. https://www.sciencedirect.com/science/article/abs/pii/S1574954122003181.
24 "Earth Species Project." Last modified 2023. Accessed June 2, 2023. https://www.earthspecies.org/.
25 "Pollinator Pathmaker." Alexandra Daisy Ginsberg. Accessed June 2, 2023. https://pollinator.art/.
26 "FluentPet: Learn to Communicate with Your Dog." Last modified 2023. Accessed June 2, 2023. https://fluent.pet/.
27 Jiabao, L., Merlin, T., Warren, C., and Montgomery, V.L. "Animal influencers as a way out of climate fatigue." *SXSW* (2023). Accessed June 2, 2023. https://schedule.sxsw.com/2023/events/PP122705.
28 Merlin Tuttle's Bat Conservation. Accessed June 2, 2023. https://www.merlintuttle.org/.
29 OctoNation. Accessed June 2, 2023. https://octonation.com/.
30 "Interview with Jacques-Yves Cousteau." *UNESCO Courier* (November 1991). Accessed June 2, 2023. https://en.unesco.org/courier/november-1991/interview-jacques-yves-cousteau-0.
31 "Bacardi Joins BCI in Saving Bats." *Bat Conservation International*. Accessed June 2, 2023. https://www.batcon.org/article/bacardi-joins-bci-in-saving-bats/.
32 "Tencent QQ." Accessed June 2, 2023. https://im.qq.com/index/.
33 "Emperor Penguins Could March to Extinction if Nations Fail to Halt Climate Change." *PBS NewsHour*. Accessed June 2, 2023. https://www.pbs.org/newshour/science/emperor-penguins-could-march-to-extinction-if-nations-fail-to-halt-climate-change.

13
DESIGNING WITH NON-HUMANS

Ralph Ghoche in Conversation With Joyce Hwang

Ralph Ghoche and Joyce Hwang

This interview took place remotely on November 12, 2021 between Ralph Ghoche, Assistant Professor of Architecture and Joyce Hwang, Associate Professor and Director of Graduate Studies at the Department of Architecture, University at Buffalo. Hwang is a registered architect in New York State who develops constructed environments that incorporate wildlife habitats. Her projects include "Bat Tower," "Bat Cloud," "Habitat Wall" and "Bower." One of her latest projects, "To Middle Species, With Love," was included in Exhibit Columbus. Past work can be viewed on Hwang's website, Ants of the Prairie.

The interview originally appeared in the following publication and has been reprinted with permission of the editors:

Journal Title: *Les Cahiers de la recherche architecturale, urbaine et paysagère/Les Cahiers: Journal for the Study of Architecture, Urbanism and Landscape*
Issue Title: L'architecture à l'épreuve de l'animal/Architectural Perspectives on the Animal
Issue Editors: Manuel Bello-Marcano, Marianne Celka and Mathias Rollot
Issue Number: 14
Date of Publication: 2022

RG: Thank you for taking the time to talk with me about your work on animal habitats. I believe it will be as new to French audiences as some of the work in France may be to people in the US.

JH: There may be more theoretical thinking about these questions in the US, but in Europe, in places like the Netherlands, Germany or the UK, it seems like there's already a lot of ongoing work. There are even building product manufacturers in Germany—such as Schwegler—that are making bricks for animals to live in. So, there's a lot of work around in Europe. You know the Mellor Primary School by Sarah Wigglesworth in the UK has a kind of insect wall as part of it. There's a very technical book that's about integrating wildlife habitat into buildings (*Designing for Biodiversity: A technical guide for new and existing buildings*) that's published by RIBA.

RG: I wondered if you could talk a little about an expression that you often use, "architect as advocate."

JH: Okay, sure. Well, I started using the phrase "architect as advocate" in thinking about some of my work when I realized that creating small scale projects that somehow drew awareness and brought interest to particular populations—whether it's bats, bees or something else—was a way that architecture could bring attention to and empathy towards particular species. And so the idea of thinking about how architects can be activists in a sense, to address situations like bird-glass collision is a big issue. Nowadays, there is an increasing number of practitioners who are working on things like bird safe building guidelines—there's even a LEED pilot credit for bird glass collision that's out there. So you can work on policy and you can advocate through policy but, as an architect and a designer, there might be ways to bring more attention to something like bird glass collision as an issue. One is just advocating for issues that impact populations that are marginalized and not thought of. When I say marginalized, I'm also thinking of humans too, so it's not just animals but how we can think of inclusion among humans in general. But in terms of thinking about non-human species, how to translate needs and so on, I don't think we, as humans are necessarily able to translate directly. It's not like we can talk to an animal and ask them questions in their own language, we can't communicate with them specifically, but there are different ways that one can observe. There are practices of observation, there's mapping. I usually do a lot of collaborative work talking with biologists and ecologists and trying to learn as much as possible. So, for example, in the most recent project that I did for an exhibit in Columbus [Indiana] called *Exhibit Columbus*—this is a minor example but it was also something that was totally visceral at the same time—we were making the structures' bases by stacking stones and creating little gaps for small amphibious animals and small terrestrial animals (Figure 13.1). While we were building the project, we saw these toads trying to jump in but it was hard for them to reach because of the height of the foundations that we had made. So, through this observation we started stacking stones around the base to make it easier for the toads to jump in. After a period of time they started inhabiting the project, even while we were designing it. It may be a ridiculous thing to talk about but I think observation is the key if you're not necessarily an expert and you're not regularly observing and researching animals, which is something we as architects are not doing.

RG: The question of representation is an important one. I think about all the philosophical and legal work that's been done in terms of giving agency to the environment, the whole concept of ecocide, to put that in relation to say, laws that are universal laws that safeguard the rights of humans. Now, of course, there is talk of the universal rights of non-humans. So, I wondered at a larger scale of the architect as advocate, do you see the role as being more all encompassing in relation to, say, non-animal actors as well? So, for example, in dealing with climatic change, to what extent does the architect have a role to play in making visible these larger atmospheric phenomena.

JH: Yeah, I think what you're saying about making things visible and representation is spot-on in terms of advocacy—not only in terms of non-human species, but climate change, biodiversity loss, carbon, and any issue that needs attention through strategies of visualization, mapping, and representation. Making projects that somehow embody the sensibilities related to issues that are important—I think that's really something that architects should aspire to do. When I think about the idea of advocacy, it's not just about making

Designing With Non-Humans: In Conversation With Joyce Hwang

FIGURE 13.1 To Middle Species With Love

visible but about being a detective as well. This goes back to observation; how can you look beyond what's already out on the table and think about what's being ignored? That could have something to do with non-human species, but it could have to do with other things too.

RG: I have a couple of questions that relate to historical examples of architects designing for animals. The one example—and I'm sure you're very familiar with it—is Berthold Lubetkin's penguin pool at London Zoo. What's interesting about it—of course you've probably heard that it's no longer used for the purposes it was designed for…

JH: Oh, yeah, I'm not surprised. I did visit it in 2000 and watched it being used, and it didn't look too comfortable.

RG: What's interesting about that project is, yes it was adapted for penguins, but mostly, it's a spectacle for humans to gawk at these penguins parading down like it's the staircase of the Paris opera house. I wondered about your work in relation to the role of human vision onto these habitats.

JH: I think there is something to be said for spectacle and its ability to bring attention. This is the power of places like zoos. One could say that it's not a good idea to keep any animals in captivity or to force animals on a ramp or to keep birds in a cage, so I completely agree with that. But I think the idea of creating a spectacle visually is certainly something that is able to draw human attention. And oftentimes, probably most people who have grown up in a city or in suburbs, their first interaction with animals is probably in a zoo. I mean mine certainly was, aside from seeing wildlife in the backyard and birds, my first fascination with animals came from going to a zoo when I was a little kid, and I'm sure that is the same with many children. Even conservation biologists use the term charisma, non-human charisma, to talk about the ability of animals that look a certain way to attract more attention, to bring attention to conserving not only that species but also other species. How can you draw from the optics and visual appearance of an animal towards conservation efforts? So, I think humans as audiences are really important. But at the same time, in my work, I'm really interested in thinking about the human audience as key. Even though the clients, or the inhabitants or users are animals, it's really important that there is a way that the project also impacts or resonates with humans as well. In the case of the Lubetkin penguin pool, from my understanding—and I'm not a historian—the distance of the pool relative to where the penguin entered the water, and the radius of the pool weren't conditions that allowed penguins to swim comfortably. When the penguins were swimming they'd have to turn in a way that was almost too soon and didn't accommodate their specific mobility preferences. In my work, paying attention to animals' preferences and tendencies is what I'm more interested in pursuing. Of course, you're familiar with the term *umwelt* by Jakob von Uexküll. I think it's really important to map out these conditions that are specific to animals and to use that information, rather than thinking of human constraints first and foremost. If you can't get a pool where penguins can swim comfortably, maybe you don't build a pool at all, and you do something else.

RG: Yeah, what happened with Lubetkin's project is interesting because he did consult with a biologist. He consulted with none other than the evolutionary biologist Julian Huxley. But despite his collaborations there were these unexpected misalignments. Architects like to talk about post-occupancy, there is a whole tradition about thinking about the way that architecture is used or misused—we can think of Bernard Tschumi—but dealing with non-humans it's particularly difficult because often we're talking about the very survival of a species. If there are misalignments, mistakes can be quite catastrophic. One can think about Biosphere 2, the closed ecological system research facility in the Arizona desert where scientists imagined they could replicate a very complex set of ecosystems. They quickly realized that, when you bring in too many variables, things get so complex that they become really hard to model, really hard to predict. So I guess I wondered about misalignments in your own work. What are some ways that your structures have been used against your intentions? Are there cases where species you hadn't considered ended up finding habitats in your work?

JH: In a project I did for Artpark, we designed a project called Bower for seven different bird species (Figure 13.2). And I consulted with a biologist to get specific recommendations for these seven bird species and it had everything to do with which direction the nesting box faced, which way the hole was going to face, either north, south, east or west, and what sizes the holes would be. I went through these specifications pretty rigorously and maintained them throughout the design. But it turned out that the bird nesting box that had a larger opening was actually not used by the bird that was intended—a Purple Martin—but was used by some other bird species that was a problem. It might have been a starling but it was being used unintentionally because of the relatively large size of the hole. And so, we were asked to put wire mesh on the holes to exclude the species they didn't want. If you look at the project it looks like the bird boxes have openings but some of the boxes are actually closed off with wire mesh behind the holes. I think another

FIGURE 13.2 Bower 2

unintended consequence happened in a project I co-directed as a faculty member where graduate students at University at Buffalo developed a design-build beehive installation. The project was essentially to move a living beehive that had formed organically behind a piece of plywood in a building that was going to be renovated. The building owner was going to give the beehive away. We, some faculty members in the architecture department, said, rather than asking a beekeeper to remove the beehive altogether, let's try to move the bees into an artificial beehive that students could design. We held a design competition for students. There were a number of different teams and four finalists. The students designed a beehive based on recommendations by the beekeeper and biologists and an ecologist who was working with Tifft Nature Preserve at the time, and is now with the Buffalo Audubon Society. And so, the students designed a beehive—they called it "Elevator B"—and built it, and it worked perfectly for a while. The beekeeper was able to move the bees from the existing hive to the student-designed beehive. The process was actually quite funny, it involved using a shopvac and sucking the bees out and capturing the queen and then blasting the bees back in the new hive. It actually worked fine. They starting building a comb, but after the winter the new beehive was not as warm as we thought it was going to be and the bees that we so carefully moved over actually died. The good news is that, after a lot of tweaking and figuring out what some of the issues were, the students and property manager built an additional layer in the beehive to maintain the thermal environment a little bit better and a new hive developed in there. This was built ten years ago and there are still bees in there right now. Another example is a project called *Bat Cloud*. I had built it for bats, but for a number of years the project had no bats in it. It got to a point where the site managers at *Tifft Nature Preserve* were thinking, well maybe it's time to take this project down because it's starting to need some touching up and because it was only intended to be a couple of years' temporary installation, and now it's been up here for five years. But as it turned out, a biologist from another university nearby, SUNY Fredonia, was doing some research looking at bat populations and actually did find bats flying in and out of this project. So now it's a strange situation because the installation was built to really just be up for a couple of years, so it's not in the best physical condition after ten years, but because of the fact that bats are in there, it can't be touched, even though it could use some repair. Yeah, misalignments happen all the time and then you sort of have to be attentive and fix things and remedy things if possible.

RG: These are good examples of what environmental philosophers call "entanglements;" the way the lines between the human/animal divide have become more or less blurred, generally towards beneficial ends. It aims to counteract problematic ideas like wilderness, the nineteenth century term for grasping nature as something completely exterior to human being. Today, you have an understanding that ecosystems pervade human spaces, that even a concrete sidewalk can have moss, grass, and insects. Your work dramatizes these ideas in very effective ways. I remember as a child we discovered a massive beehive hanging from the eaves of our roof and we had an exterminator take it down. We took it apart and learned that the beehive was composed of fiberglass insulation from the roof that the bees had been poaching. So, they actually made their natural/unnatural habitat. Of course, as the globe gets increasingly urbanized, we may have no choice but to consider these compromised conditions. In your bat habitats for Griffis Sculpture Park, you talk about how you could have purchased bat habitats that blend into the environment

FIGURE 13.3   Bat Tower

but yours intentionally stand out (Figure 13.3). You're very much conscious about the fact that these are artifices, that these are constructed environments. You make that a key element in your work.

JH: I think that's almost a given for most architects. I can't imagine that any architect is thinking that what they are building is natural, even if what you're using are natural materials. But the question of constructing artifice as a form of habitat is certainly not new and it's hard to try to think otherwise because the world we're living in right now, a large part of the world is urbanized and the rate of urbanization is only happening faster and urbanization is one of the key causes of biodiversity loss and this is something that is urgent. We, as architects and humans, have to deal with this. So, I think that intentionally making a structure conspicuous and very markedly part of the designed environment is part of the agenda. I don't think it's helpful at all to think that only planting trees and including constructions that are not artificial, that that's the only way to make any difference. The fact is, what we really need to address is how to urbanize while increasing biodiversity.

RG: Is it fair to say that your clients are animals and insects while your audiences are humans? In the work you currently have up in Columbus you introduce what look like historical elements in your work. They remind me of Constructivist forms, I'm thinking Vladimir Tatlin. I wondered if that was intentional, and I wondered how you navigate the many audiences looking at your work, which includes architects seeing your work not just on site, but also in architectural journals.

JH: First, I think it is right to say that my clients, or the users, or the occupants or the stakeholders of the projects are animals and that the audiences are humans. But ultimately, humans are stakeholders as well. I think with every project it's quite different. For Exhibit Columbus, the audience is not just the people in Columbus because Columbus is a small city—there's around 50,000 people—it's an incredible town that has an incredible legacy of architecture. Any random person that you run into on the street knows a lot about Saarinen or architects from the twentieth century because of the legacy of architecture in Columbus and the Cummins Foundation Program. There's a particular local audience that's really in tune with architects, so one of the intentions of that particular program was that the installation somehow resonate in some way with the site. The sites are designed by specific architects, so in my case, our project was in Mill Race Park, which is a park that was designed by Michael Van Valkenburgh and the structures in that park were designed by Stanley Saitowitz. So, there's definitely a clear reference between some of the structures that I built and the observation tower on site, which is a concrete, almost Brutalist-looking tower. It's interesting that you mention the Constructivists but I didn't have that in mind. In fact, the bat houses are modeled after a very conventional bat house typology which is the Rocketbox bat house. That's a type of bat house that has been particularly helpful for the endangered Indiana bat. So, I was looking at bat house typologies and trying to find a way to enhance them with other sorts of habitat conducive structures.

RG: I think somewhere you talk about your work as challenging the modern values of cleanliness, one could also add hygiene, the whole modern rhetoric of the need for spaces to have abundant light and air. When you think about your pieces also participating in the world of human architecture, do they suggest something about how architecture itself or the design of buildings might move towards the creation of more ecosystem-rich urban environments?

JH: In terms of challenging ideas of cleanliness and hygiene, I think certainly one of the ways that my work challenges that paradigm is that a lot of the projects I've designed are considered to be sculptures or art work and belong in a sculpture park. And because of the fact that they are intended to be inhabited by animals, you do see a level of occupation that then defines what the project is. Unlike typical sculptures and artwork, the installations shouldn't be repaired and cleaned regularly, as cleaning regularly would be detrimental to the habitat. So, the Bat Tower in Griffis Sculpture Park is one where, when you go and look at it right now, there is actually bat shit everywhere, bat guano sitting around on the structure itself. Thinking about how built structures can be part of the ecosystem, part of the environment and don't necessarily need consistent maintenance and cleaning is something that interests me. I bought a vacant property in Buffalo recently and I'm going to be, hopefully in the near future, designing a house and studio for myself there. So, I'm thinking of integrating a lot of these ideas into that project, so then we'll really see how much I can withstand not cleaning things.

RG: Your work makes me think of François Roche's (R&Sie(n)) *I'm Lost in Paris* house where there is a codependence established between the user, the client, and the natural organisms. In the case of that particular house, it finds itself in the middle of a courtyard and is incredibly exposed because you have the windows of the adjoining apartment building looking onto it. The only way that the client can have any privacy, therefore, is to maintain this rich foliage that is composed of a specific species of fern that need to be fed a particular enzyme in each of the hundreds of watering flasks holding the plants.

So, we get back to the idea of entanglements, the human, in order to get privacy—which is a very *human* need—has to enter into a relationship based on interdependencies with a non-human species. I wonder, because you mention designing a house, so we see the habitats giving animals something but is there a next step in your work to move towards a mutual relationship.

JH: I think that there already is a mutual relationship between humans and animals that we just have to recognize. One of the reasons that people are trying to attract bats more and more these days—which was not the case ten years ago when I first moved to Buffalo—the reason people are wanting to bring bats back is that they eat a lot of mosquitos. Informally, at a party I've heard people say, "Oh, I'm building some bat houses because I can't deal with these mosquitos anymore." So, I think there already is a co-dependency or mutual relationship to some extent. But in terms of upcoming house project, I was thinking about those birds that build their nests with trash, or birds that build their nests with cigarette butts, and they found that cigarette butts are really useful in terms of keeping certain parasites away from nests, and some birds will even seek out cigarette butts to build their nests. Part of me was wondering if there are certain ways that animals building nests or animals building something for themselves would also be a way to start contributing to the larger home environment. If you had a wall that was filled with bird nests, and let's say that the birds all used cigarette butts, would that be a way to prevent parasites from growing in your wall. I don't know, I'm not projecting that specific scenario is going to happen, but I think there's always going to be some kind of mutual relationship regardless what happens.

RG: In thinking about historical example where architects have designed for animals or non-human species, one really interesting one is the cowshed at Gut Garkau near Lübeck, Germany by the architect Hugo Häring. It was designed for animals in a similar way as people were designing kitchens through understanding human behaviors or applying Taylorist models to production to streamline production lines. There is real attentiveness to the way that cows interact, the way they feed, the way they bully each other. So, one element that often gets highlighted is the way there are no sharp corners, all the corners are rounded, and that's because some of the older cows would push calves into corners, potentially injuring them. It's an interesting project because to me, it's always seemed as the degree zero of functionalism. It's of course ironic that the most functionalist building happens also not to be intended for humans, but for cows. In a way, there is something very similar in your process— you are also working with biologists and ecologists, you are trying to really understand the behaviors, the actions, the cycles that these animals engage in. I wondered if you could take us through your design process and tell us a little about how the forms of your project emerge.

JH: I actually started thinking about Temple Grandin's humane slaughter machine when you described the cowshed. In terms of my process, I think there's certainly a lot that starts with research, talking with biologists, and so on. But there's also a big part that starts with trying to distill spatial conditions from research. This is something I do with my students a lot where we will look at a particular animal or a particular species and we'll go through an exercise where we talk about this notion of *umwelt*, a term coined by Jakob von Uexküll to talk about the specific environment of species. Of course, the example he offers that is kind of known is the tick and the deer. Even though the tick and the deer are both in the forest, the tick's specific environment is the deer. The tick doesn't care about the forest while the deer cares about the forest. The thing that I always try to do first is, after doing a whole bunch of research and talking to people and making observations, to really try to think about what the specific conditions are to make a set of spatial types or spatial conditions that I can start to work with. So, then I'll work on a process where I'll say, "ok, I am going to work with slotted spaces that are roughly ¾ of an inch thick and I'm going to work on surfaces that have ridging." That's a specific thing related to bats, bats climbing into spaces and the kinds of gaps that bats will occupy. So, I'll work with that as a kind of vocabulary, let's say, and then start building models, say physical models, Rhino models, or whatever, using these elements and bringing them together and thinking about that as spatial building blocks. And then I'll usually work through a lot of different iterations and bring the biologist back into the picture. One thing I found is, working with people who are not architects it's helpful for them to look at physical models. Often what's helpful is if I drag a big physical model with me, even if I'm going out for a coffee with a biologist, I'll put a physical model in my arm and say what do you think of this? So, it's a back and forth in that way. In a project like Exhibit Columbus it's slightly different. I started out with a typical bat house, I started out with the Rocketbox bat house which has already been something that's used, that's been understood to work for Indiana bats. It has a specific set of dimensions. So, I literally just took the dimensions that were provided from the off-the-shelf DIY bat houses and I used that as a starting point to then build off of.

RG: And do you put some of these models in the field? Are there ways to test them out? I guess you have done a number of bat houses, what have you learned from the various iterations?

JH: Well for Bat Habitat Wall Cloud, for the first version, we made some prototypes and put them outside just to see what would happen. We weren't testing necessarily for bat occupation but we were looking at temperature monitoring, things like that. So, I had a thermal camera that I was using to take photos of the bat house. One thing we were also trying to look at was how well heat would be retained inside the bat house pods. We were doing experiments like heating up gel packs to a degree where they would be thermally equivalent to a warm blooded creature and we were shoving them into the pods to see how long the heat would stay. Looking at thermal environments is something we'll experiment with too.

RG: One could argue that architects have no real center to their discipline, they don't necessarily have a real source of expertise. They don't build buildings, they draw drawings that in turn prescribe buildings to be built by others. And that might be an advantage to the field because, if architects are experts at anything, it is at bringing multiple disciplines in conversation with each other. You've talked a lot about your collaborations with biologists and I'm wondering if you think there might be a model there in your relationships with biologists that could be expanded?

JH: In earlier projects, I was collaborating with biologists and ecologists and asking for their advice as consultants, in a way. My later projects like Life Support, for example, which was completed in 2019, was a collaboration with an ecologist in Australia, Darren LeRoux, but also a data visualization designer, Mitchell Whitelaw, who teaches in the school of Arts and Design in Australian National University. Interestingly, he and I were at the same conference together with Darren who was presenting his research on very large trees and their ecological benefit. And it was Mitchell who thought that the three of us could actually work together. At that point I wasn't even thinking, "oh, I'm going to reach out to this data visualization guy and try to collaborate" but he saw the potential in it. So, the three of us did work collectively on a project in Australia. I was doing most of the architectural design and collaborating with a structural engineer. The project was basically to use a four-hundred-year-old yellowbox eucalyptus tree that was slated to be removed from a residential neighborhood in Canberra. And normally when trees are taken down from residential neighborhoods in Australia, which is quite common apparently because of the heaviness of the branches and the fact that they frequently drop and damage property. But it's a problem to just cut it down and make it into firewood because large, old trees inherently have a lot of ecological value. The knot holes, peeling bark, and everything about the tree itself provides rich ecological habitats. And so, I worked with these two collaborators in Australia, first to go through the process of figuring out how we would take down the tree and cut it in as few pieces as possible so that it could be used in a sculptural design project. And once the tree was taken down, we had to develop a digital model of the existing tree parts, to even begin to work with it. There was a lot of work with the data visualization designer to use photogrammetry to try to capture the form of the tree, and later on he became very instrumental in putting camera traps on the installation documenting species that were coming to the area around the project.

RG: Were the cameras accessible to watch online?

JH: Yes, Mitchell, the digital designer made a website called Molonglo Life and that website hosted a lot of videos and photos that were taken from the camera traps. And the hope was that people would start to identify species almost as a form of citizen science. In more recent collaboration for our Exhibit Columbus Project, I collaborated with biologists from the Indiana Department of Natural Resources, and have also been working with two musicians who have been experimenting with bat sounds. We installed ultrasonic bat detectors that are recording bat sounds in the area. We've been collecting these sounds and giving them to musicians to use as source material for making compositions. It's actually really cool, this is the thing I'm really excited about right now. These musicians are basically exploring and mixing bat sounds and it almost sounds like Berlin techno. Even for the opening of Exhibit Columbus we had a bat concert where musicians were taking the bat sounds and playing around with them and making soundtracks with them. So, I think there are certainly collaborations in trying to realize a project but there is also collaboration in what a project can produce, what the afterlife of the experience of the project can be.

RG: My last question goes back to the idea of translating back to human users from animal users. Bats have very specific needs. Human needs, however, are a little bit more difficult to pin down. There are all sorts of needs that are not material needs that humans have. But when focusing on animals and their narrower material needs, I'm wondering if your research has given you new perspective on the larger set of needs that humans have. While there are problems with translating certain lessons from the animal kingdom to humans, I wonder if there are also opportunities in considering what humans can learn from ways that animals respond to built environments.

JH: I don't think I've been thinking along those lines of how do we take lessons from what we've learned about animals as inhabitants or stakeholders in the world and how we translate those as humans in the world. I don't think about it so explicitly as a translation, but I think that there is a lot about what we learn from the way animals occupy the world that affect the way humans occupy the world and should become part of our ways of understanding things. So, for example, the fact that we understand what a setback is based on sunlight. That's something that feels almost normal now. We understand that if a building is tall enough, there is a need for sunlight to come in. What we don't think of are setbacks for things like squirrels jumping. A squirrel jumping setback should be around 10 feet. So, if you don't want a squirrel to jump into your house from a tree, you better set your house 10 feet away from a tree. So, I think there are certain things about animal behavior that could work their way into the way we think about our designed environments. That's something I think about a lot when I think about both animal space and human space.

14

THE PROBLEM IS THE BURNING HOUSE

Catherine Page Harris

Carey Wolfe in *Animal Rites* (2003) cites Tim Luke's condor thought experiment. Imagine a burning house with a baby human and a baby condor in it with the requirement that the listener choose one baby to save. As humans, we will not imagine leaving the human baby to die, but the condor is more valuable than the baby if measured as a ratio of population. One condor chick is 1/300 of the total living population of condors. One human baby is 1/7,880,000,000 of the total human population.[1] Perhaps Luke constructs this paradox echoing Robinson Jeffers' 1932 poem "Hurt Hawks": "*I'd sooner, except the penalties, kill a man than a hawk*;" an odd place to start for this chapter given that even the famously misanthropist Edward Abbey finds Jeffers too dour for his taste.[2] It is important to recognize in this conversation the dangers of misanthropy – the racist, eugenicist past of environmentalism, which haunts a conversation which must embrace indigenous rights, black and brown rights, racial and cultural equity among all humans. Developing an "and" rather than a zero-sum game, however, will offer sovereignty to human and non-human, recognizing the necessity of valuing different lived experiences for humans, but knowing that genetically humans are perhaps related through a single common ancestor as recently as 2000 years ago.[3]

Embedded in the condor paradox are questions of value, biodiversity, and survival. The paradox is not a paradox if the reader resists the premise that survival is a zero-sum game and instead imagines that the listener's role is to douse the burning house rather than run out with one baby or another. The paradox is resolved if I understand the planetary experience to be commensal. The first uses of the word commensal described those who eat together. The word now describes biological life that shares resources and, in particular, the micro-biotic life that lives in human mucus. We humans and non-humans are in a commensal community at many scales, and humans need to reframe their designs to value all lives in that community. There is no putting out the fire, only adapting to it. Part of the Poe-esque horror of our predicament is that we left the burner on, we left the heater in front of the drapes, we set the house on fire. Can we resolve the condor paradox by realizing that the problem is the burning house, not the species of the infant? Three writers offer concepts to find our way into this design practice: "Full responsibility" and "unconditional hospitality" remove profit/capital as the criteria for resource decisions.[4] "A heterogenous multiplicity of the living" creates space for all beings to have rights.[5] "Intensive spaces of becoming" require and create habitat for all beings.[6]

DOI: 10.4324/9781003403494-17

Section One: Post-humanist Philosophy

Post-humanist philosophy offers terms to guide this exploration. To clarify, post-humanism is not trans-humanism, the pursuit of improving human existence through cyborg and other means, but rather the philosophical pursuit of the proposition that current research has led us to a point where non-human sentience is similar and equal in many ways to human sentience and that a human-first stance is tending towards a ruined planet and lessened life quality for all species. Post-humanism follows on many theories that have sought to redefine the relationship between humans and non-humans as equal, not hierarchical, poet Jeffers' "inhumanism" being an example from the early 20th century.[7] In *Before the Law* (2013), Wolfe looks at the legal ramifications of such an understanding.[8] His conclusion works with what might be called the biodiversity paradox for humans, which he points out a decade earlier in *Animal Rites*, 2003.[9] Quoting Tim Luke from his 1988 article "Dreams of Deep Ecology" in *Telos*, Wolfe discusses the deep ecology paradox of placing value on the life of a human child (a member of a class of a species whose numbers have grown exponentially) vs the life of a condor hatchling (a member of a class of species whose numbers have been down to fewer than 460 in any given year since 1967:

> And for the very same reasons, an ethics of pure equilibrium without decision, without discrimination—without, in short, selection and a perspective—would be, paradoxically, unethical. It's not that we shouldn't strive for unconditional hospitality and endeavor to be fully responsible; it's simply that to do so, it is necessary to do so selectively and partially, thus conditionally, which in turn calls forth the need to be more fully responsible than we have already been.[10]

From Wolfe comes "unconditional hospitality" and "full responsibility" as frameworks for physical manifestations and design decisions.

Turning to another post-humanist writer, Rosi Braidotti, in *The Posthuman*, 2013, we find a philosophical stance of constitution championed by several other writers, including J. B. MacKinnon in *The Once and Future World*, of the same year. The existence of non-human sentience provides a collaboration with human sentience that is constitutive and thus, in part or sum, creates human consciousness and experience. (Or perhaps to reach even further back, from Gertrude Stein, "I am I because my little dog knows me..."[11]) This line of thinking suggests that it is not only good manners (Wolfe's unconditional hospitality) to welcome non-human sentience into our world, but extending that hospitality, in fact, creates human sentience. Similar to the concept of terroir, a landscape creates the biota living in and on it. Rosi Braidotti posits in her 2013 *The Posthuman*:

> In my view, the point about post human relations, however, is to see the inter-relation human/animal as constitutive of the identity of each. It is a transformative or symbiotic relation that hybridizes and alters the 'nature' of each one and foregrounds the middle grounds of their interaction. This is the *milieu* of the human/non-human continuum and it needs to be explored as an open experiment, not as a foregone moral conclusion about allegedly universal values or qualities.... Intensive spaces of becoming have to be opened and more importantly, to be kept open.[12]

This term, "intensive spaces of becoming" can also be applied to a physical design decision-making process. How can landscape provide opportunities for this constitutive identity located not only in the domestic sphere as Haraway does, but also in the larger public sphere?

Jacques Derrida, in his ten-hour 1997 address to the Cerisy, France conference, "The Autobiographical Animal," published in 2009 as *The Animal That Therefore I Am*, by Fordham Press, develops a term that may also be relevant to the pursuit of design in a post-humanist world, "a heterogeneous multiplicity of the living."[13] In his talk, Derrida also, notably, protests the treatment of industrialized animal agriculture and production. Derrida refers to these practices as horrific and goes on – "and all of that in the service of a certain being and the so-called human well-being of man."[14] His qualification of these practices as being in that service indicates that Derrida doubts that the industrial treatment of animals for food consumption and production contributes to the well-being of humans.[15] I extrapolate from that an interest in the Haraway, Braidotti, and MacKinnon position that non-human and human together create constitutive identities. In his statement of his thesis, Derrida articulates "1. This abyssal rupture doesn't describe two edges, a unilinear and indivisible line having two edges, Man and Animal, in general," which is a liminal area where human and animal are not easily divided.[16] From Derrida, I will add the phrase, "a heterogeneous multiplicity of the living," as a typology for the post-human landscape and reference the "line having two edges."

I propose meeting Wolfe, Braidotti, and Derrida's proposals with typologies for landscape design – the thin line with two edges, the smooth surface, and layered multiple connections. These design techniques are derived from landscape architecture's engagement with a thickened space at the surface of the planet. A landscape is chthonic and microbial; it is the movement of the soles of our feet on the surface of the ground, migrating, living, negotiating from Africa to Arctic. Landscape is the envelopes of air around human and animal movement. This space of design asserts that there are joint accommodations being made by commensalist species – species who share resources – a phrase that began with the idea of sharing food. Humans can develop further animal bridges and continue backyard habitat networks. Non-humans continue to adapt and change in response to urban space configurations. Pigeons, eagles, falcons, coyotes, cockroaches, mosquitoes, and their kin form commensalist networks with humans at many scales.

This section investigates the possible configurations that may be collaborations among human and non-human in a series of proposal drawings for shared landscapes.

Section Two: Typologies

The Thin Line with Two Edges: Example

The thin line with two edges visibly exists in the efforts of Cape Cod beaches to restore piping plover populations (*Charadrius melodus*). In the four decades since populations of piping plovers across North America were federally listed as endangered, management plans have become community projects. Plover beach restrictions have changed the use of beaches through education, enforcement, and a literal thin line of string. Bolstered by signage and regulation at the beach parking lot, ultimately the birds are protected not by barriers, which would defeat the purpose of their nesting habitat, but by a simple public agreement of lines of usage. Observed in summer 2022 at Head of the Meadow's beach, long-time visitors talked about the differences in beach grass cover, the bird population growth, and the beauty of the beach, all the while observing the restriction delineated, but not imposed, by a string. A string, of course, cannot stop a human, except by agreement. As shown in the two photo collages below, while birds visit outside the string, humans align themselves along the crest of the beach high water line and leave the

FIGURE 14.1 The Thin Line

back beach for bird nesting. An extraordinary piece of shared human and non-human spatial landscape design, accomplished with the mildest of methods (Figure 14.1).

On the left, the string and the sand surface beneath, showing the difference in tread, in human use and bird use. On the right, the actual agent of change, a wooden post with string shown dividing beach umbrellas from nesting habitat. In the middle, an encounter of beachboarding child and plover resting after a swooping dive across the waves. The plover is on the human side of the thin line, but with each side sharing one ecosystem, with two habitats defined by the thin line. Within this line sits education of the public, a sense of community bonding over the protection of the birds and unabashed love for fellow non-human inhabitants.

There are parking lot admonitions and instructions. There is a literal carpet, blue not red, that marks the linear approach to the beach and prevents the highest foot traffic area from eroding beach sands. The pathway's existence, more than the strings' admonition, keeps feet off the protected area. The beach line is a literal thin line governing human behavior for the benefit of non-human habitat. Derrida describes an abyss between Man and Animal, which he argues has been bolstered by language as a criteria, and which he rejects, describing instead a thin line with animal and man that joins at its threshold.[17] This thin line is not an abyss, but a joint.

For urban landscapes, I propose a thin line akin to Derrida's but one which perhaps involves gaps through which animal populations can move, perforations where human and non-human accommodate one another, a weaving with the surface. Derrida's line has two edges, dividing a continuum of "human and animal" (human and non-human, human and more than human) and I see those two edges slipped, shifted, to create space for all of us multi-species commensal creatures.

The thin line represents an opening of spaces within human urban and non-urban fabrics. Perhaps also an articulation of Wolfe's "unconditional hospitality" the thin line can be reframed as accommodation – notably the provision of passage through, under, above, lines drawn in the landscape. Wilderness protections have begun with animal overpasses in Banff, Canada and connecting northern parks in the United States. Perhaps it is time to imagine extending this paradigm to all road developments. Human long-distance travel can move to electric railroads, narrower impositions on the landscape for the number of passengers. If those tracks were designed with a vertical weave, sometimes above, sometimes below, and sometimes on the ground, then they could be major people movers rather than ten-lane highways that effectively cut off one portion of a landscape from another.

This thin line can also be seed-established landscapes – one of the largest contributors to the carbon footprint of construction is the movement of materials to the site. What prevents landscapes from being transported to the site as seeds? A landscape could fit in a bag. Turtle Island Restoration in California is aiming to plant 10,000 redwood trees starting from seeds that are the size of a tomato seed. Humans are impatient, not quite six foot tall, and short-lived, but a redwood grows 150 feet in our life span. Seeds can start grasslands or meadows in a design, and often do, but what if landscape architecture used largely direct seeding techniques? This method would prevent plantings that require heavy care as well. A seed won't germinate where it can't survive. This pushes directly against the capital structure of landscape design from the nurseries that support the industry with contract grown plants for large scale jobs and seasonal color for residential clients. It would also liberate plants from root bondage that must surely occur in pots and burlap bags. More specifically, it would remove fossil fuels from the production of planting by removing long-term fertilizing and trucking full-grown trees and shrubs hundreds of miles.

Philosophically, the seeded landscape is a slipped thin line, a space of perforation. Seeding allows a design to be initiated by human concerns – providing shade, a vista, a beautiful flower to contemplate, or food – and still allows the landscape to be determined by its own conditions and capacities.

The Smooth Surface

Derrida's "heterogenous multiplicity of the living" is a smooth surface between animal and man. Here the smooth surface can be typologically represented as a field of opportunities equally spread. This can be seen in Wolfe's "full responsibility" as a typology – a framework for understanding a decision-making process that includes Derrida's "heterogenous multiplicity of the living" and risks to those living beings. It is a drawing of a system of internalizing rather than externalizing potential costs to life and welfare of all living beings. It assumes, as a typology, that humans are the actors and generators of the design, and that we are responsible to the living. It is also governed by Wolfe's concept of "conditional."

Smooth surfaces in landscapes could be manifested in many ways. A literal smooth surface could be adopting the rapidly advancing techniques of no-till or regenerative agriculture. No-till is a designation for fields that are not plowed. Carbon quantification supports these no-till farming practices I first experienced in Marin County in 1994. In 2022, from Australia to the IPCC, voices are saying that conventional tilling of the soil releases carbon into the atmosphere at rates we can no longer afford. A no-till landscape reduces excavation to a minimum. Rather than scraping the surface of a site to remove and replace topsoil, mulching, stacking, piling would be actions of design. Design would be more additive than subtractive. A no-till landscape would preserve the carbon sequestration of soils. Instead of balancing cut and fill, no-till would balance carbon release and sequestration, with the weight heavily on the side of sequestration. Practices landscape architecture could borrow from regenerative agriculture include targeted use of goats or sheep for vegetation control, sheet mulching and interplanting as seasons change, and reducing or eliminating fossil fuel-derived fertilizers and pesticides. There can be no "multiplicity of the living" if the materials of our landscape actively poison the biota in them.

Braidotti's "open ended space of becoming" as a typology creates maximum opportunities for interaction between human and non-human species. Similar to the "thin line," this also generates the possibility of change as befits the constitutive quality of her proposal. Like a mosaic drawing of vegetation types, this typology creates complex edges to facilitate multiple interactions and allow for growth and change of the individual elements.

The road, the highway, the country road, the suburban drive, and the urban grid, all create spaces throughout human occupation of the landscape. Derrick Jensen refers to cars as killing machines, and certainly, as elucidated above, roads themselves are the agents facilitating roadkill, splitting habitats, and spreading human influence into as yet unbuilt areas. The 2001 Roadless Rule "establishes prohibitions on road construction, road reconstruction, and timber harvesting on 58.5 million acres of inventoried roadless areas on National Forest System lands. The intent of the 2001 Roadless Rule is to provide lasting protection for inventoried roadless areas within the National Forest System in the context of multiple-use management." Beyond Forest Service lands, roads are sites of contention among human and non-human use.

Applying typologies, a road of "unconditional hospitality" finds the topographic ability to be sunk underground or lifted above ground to allow the passage of prong horn antelope across the road.

Slow Spaces

"Full responsibility" requires slow spaces. In the areas where the road is above ground, it is slower and more resistant to the automobile's ability to move too fast to see the bear, elk, moose, antelope, or opossum in time to avoid collision. These slow spaces will require humans to notice in more detail than the current "deer crossing" signs. The road, narrowed, will require slower driving. "Open spaces of becoming" in a road will be migrations, paths for humans to join in the great flows of non-human motion through the patterns of resources and seasons. Roads will become paths for gathering, nodal spaces, where the endless vistas of linear lines, will become influenced by the topography and form narrow tracks, and then widen out into gathering spaces for multi-species.

A landscape of "unconditional hospitality" is a place shared resources limit on users. A park might be a space where migrating animals can forage, where humans experience the foraging of non-humans and forage themselves. Landscapes can become spaces where humans become educated about the native plants edible to them. As "open ended spaces of becoming" a park can be places where humans can slow down with wild turkeys or mule deer and create a place for themselves in the hierarchy of the flock or pack, as contemporary writer Joe Hutto does. These spaces can create a family of non-humans and humans, to live along side, and experience the world by taste and foraging. These spaces for foraging become transformational as places where paying attention to the vegetation and the non-human, collaborate to create an intensive space of becoming for humans, and perhaps a space where non-humans transform as well.

Policy plays a role in designing space. I will end here with posing these three shifts in human-non-human relationships to answer Jeffers' "recognition of transhuman magnificence."

1 *unconditional hospitality* removes profit/capital as the criteria for resource decisions
2 *a heterogenous multiplicity of the living* creates space for all beings to have rights
3 *intensive spaces of becoming* require habitats for all beings

Section Three: Commensalist Projects

It began with a horse in a field in Marfa, Texas. I was looking for a physical metaphor for sharing the planet's resources and I wondered what it would do to share a meal with that horse. For several years later, I was building tables and sharing meals with horses, cows, chickens, turkeys, goats, bees, in two continents. Some tables were site specific, like the month-long project for the Hansthom summer LandShape Festival, where my family and I traveled to different farms in North Jutland, Denmark, worked on the farms, made tables from materials available – weaving willows, constructing with wood slabs, abandoned renovation leftovers – and shared meals with the animals of the farm. Or the marble table installed permanently at the Marble House Residency in Vermont, where we shared a meal with bees and ate surrounded by Vermont's summer largess of flowers and wild herbs on a slab of marble from a debris pile transformed by farm post legs and a fancy repast. Some tables were recycled cardboard modular pieces that grew and shifted with the needs of the space and the animals like the three-month exhibit at the Center for Contemporary Arts in Santa Fe where I made public events sharing meals with bees, turkeys pardoned from Thanksgiving tables, and goats to bring in the new year.

This prop, the shared meal and the table, allowed transpecies experiences that embodied the communication I was seeking. In Cucamonga, CA, I was invited to install the cardboard table version and do a performance. I found a friend with chickens and hung out with the chickens for a day before taking Henrietta chicken to the event. Henrietta and I shared watermelon, perfect for the heat in inland southern California. Initially she stayed on her side of the 4' x 4' square laminated cardboard table. Then, as our hour progressed, she came to the center and we bobbed our heads, shifted eye gazes and ate from the same pile of watermelon cubes. We began to copy each other's movements, communicating with gesture. She had a low chortle which I could mimic, thus we conversed some verbally, but mostly the movements of her body began to imprint on mine. It felt like we had locked into a space together, where her evolution as a tiny dinosaur and mine as a primate, as well as her species domestication and my species interweaving into that domestication, all came together in sharing a bubble of time in a crowded gallery with photographs being taken and students and members of the public milling through. After eating with Henrietta for an hour, I went outside and found that a mixed and merry company of students was sitting enjoying the fruit I had put out on an outside extension of the cardboard modular table. I sat with them for a while, enjoying their jokes and repartee. I felt I could cross over and see them mimicking movements and sounds. I was hearing the human – human version of my human – non-human conversation (Figure 14.2).

Sharing a Drink

Sharing a Drink (ongoing, begun in 2017) takes a series of installations both inside and outside a gallery with a wildlife camera and set ups for humans and non-humans to drink. In the wildlife refuge, I use established drinkers, and in the gallery, I set up conditions for people (and others) to drink water. These waters are locally harvested (from rivers or stores) and always provided with filters and instructions. I pull video from both the outdoor installations and from the indoor ones and repurpose that video in each installation. Shown here is the installation at the Rainosek Gallery in Albuquerque, NM (2018) not pictured, at Kunst(Zeug)Haus in Zurich, Switzerland (2020) and the Santa Fe Art Institute (2019).

FIGURE 14.2 Transpecies Repast

Conclusion

Thin lines, seeded spaces, smooth surfaces, and hospitality lead to a landscape architecture around the burning house. They are the incremental changes that lead to the responsibility of practice. Our house is burning; this is not a theoretical question. We do not now save all the children, human or more than human. Our practice can shift and with that shift of practice, our minds creep into the radical consciousness of sharing commensalist spaces. With those spaces come caring and relational landscapes. If our allowed actions are freed from the heroism of saviors snatching life from the jaws of fire, we can behave through practice and thin lines, responsibility and smooth surfaces, in our constituted identities created by the sentient world.

Notes

1 Cary Wolfe, *Animal Rites: American Culture, the Discourse of the Species and Posthumanist Theory* (Chicago: University of Chicago Press, 2003), 26.
2 "Hurt Hawks by Robinson Jeffers," Poetry Foundation, accessed May 16, 2023, https://www.poetryfoundation.org/poems/51675/hurt-hawks.
3 Scott Hershberger, "Humans Are All More Closely Related than We Commonly Think," *Scientific American*, October 5, 2020, https://www.scientificamerican.com/article/humans-are-all-more-closely-related-than-we-commonly-think/.

4 Cary Wolfe, *Before the Law: Humans and Other Animals in a Biopolitical Frame* (Chicago: University of Chicago Press, 2013), 86, 92–93.
5 Jacques Derrida, *The Animal that Therefore I am* (New York: Fordham University Press, 2008), 31.
6 Rosi Braidotti, *The Posthuman* (Hoboken: Wiley, 2013), 80.
7 "Robinson Jeffers," Poetry Foundation, accessed January 3, 2023, https://www.poetryfoundation.org/poets/robinson-jeffers. Robinson Jeffers "inhumanism," which he explained was "a shifting of emphasis from man to not man; the rejection of human solipsism and recognition of the transhuman magnificence. ... It offers a reasonable detachment as a rule of conduct, instead of love, hate, and envy."
8 Wolfe, *Before the Law*, 58–64, 85.
9 Wolfe, *Animal Rites*, 23–24.
10 Wolfe, *Before the Law*, 86.
11 Gertrude Stein, *What are Masterpieces?* (New York: Pitman Pub. Corp, 1970), 20. Accessed on Internet Archive.
12 Braidotti, *The Posthuman*, 79–80.
13 Derrida, *The Animal*, 31.
14 Derrida, *The Animal*.
15 Derrida, *The Animal*, 26.
16 Derrida, *The Animal*, 31.
17 Derrida, *The Animal*, 19–21.

15

EVERYTHING WITH WINGS

Sarah Walko and Gabriel Willow

Unfolding Process

I have a process of traversing landscapes, collecting materials to work with in the studio. Mostly, the landscapes are forests in the northeast, but when visiting anywhere, I take the time to explore the environment and collect. As I walk, I am focused on deeply looking and listening. I cross paths with animals, both living and dead. I try not to startle or interfere with the activities of the living ones. A deer will immediately turn towards a new direction, swallowed by the density of the trees within moments. A fox will look me straight in the eyes and seem bothered but continue on their path, aware of me but unthreatened, granting me more extensive time observing them. When I come across animals who are no longer alive I thank them for their contribution to the story of the world, and for disintegrating back into the forest floor, enriching the soil for new growth.

I witness the slow rebirth of spring, noting which plants sprout first, just as in the fall I watched the slow refolding of growth in preparation for winter, the forest browning and becoming barren or blanketed with snow. I see the insect's parade and listen to the birds of each season. At dusk, I may see a great blue heron land in shallow streams and feast on fish. Each place has its own active conversation happening, written in the stones, the paths, the cliffs, the trees, the plants, the leaves. Author Sophie Strand has said, "I strongly believe that all thinking happens interstitially – between beings, ideas, differences, mythical gradients." I believe all creation does too. This experience of collecting is vital to the finished pieces of my sculptures.

Friends and acquaintances also collect materials for me from all over the world. One afternoon I arrived home to a box of camel bones, found in The Empty Quarter desert in Saudi Arabia, their surfaces perfectly clean from the wind, sand, and sun exposure. A package arrived filled with bird's nest coral from the Indian Ocean, and another arrived containing a plump abandoned hornet's nest, an attached note from the sender about ascending a very tall ladder to pull it from the tree. These are artifacts of lives lived and the stories are held within the remaining objects, stories that have also traveled through a life of a being, or many, and then from hand to hand, land to land.

In the studio branches and bark, nests and bones, feathers and wings, seed pods and leaves join tables filled with other materials; beads, pins, rhinestones, found objects, test tubes,

DOI: 10.4324/9781003403494-18

FIGURE 15.1   Everything With Wings, 2020, Sarah Walko

microscope slides, beakers, Petri dishes. And they then come together into sculptures of an embellished hybrid language of nature and culture. The ephemera from the natural world meet antiquities of mystical science and artifacts from culture to tell their own stories. In my series, Glass Orchestras, I combine these objects into small sculptures and insert each one into a test tube. The final installations of these pieces are organized with hundreds of test tube sculptures arranged on a wall and become less about each of the individual objects themselves and more about the relationships between one another, creating a kind of language or new story using the pieces and parts of a former one (Figure 15.1).

I'm interested in creating a mythological language at the intersection of nature and culture. I'm interested in creating work that allows perceptions to shift from the historical to the narrative, the scientific to the personal, the common place to the magical. The work explores shifting the relationship we have between species from one that is hierarchical to one that is a network, interconnected and diverse where we can all flourish. The work aims at also capturing/creating a sense of wonder and awe, creating a space where we can witness the incredible stories of the land and all beings around us but also feel each of us is included.

The term Bardo is used in Tibetan Buddhism, and it refers to the liminal realm between death and rebirth. It is the place where a soul confronts and works through issues before entering into a new vessel, container, or body. This liminal space is what art can provide for us as we confront and work through issues before entering into a new chapter. And in this space, we can look outside the bounds of human culture and human-centered narratives for our new shape, for what we could become.

Using materials collected in landscapes as well as found objects that are repurposed, my work is about creating a mythological narrative that doesn't aim to preserve nature but celebrates, elevates, and transforms it while referencing ceremonies and rituals. The materials honor the stories of the earth they each carry and the repurposed materials create art out of objects made to simply be discarded after use. The works also specifically incorporate materials that were historically often considered low art associated with traditional feminine handicrafts. Carrying the wisdom of nature, I make feminist-futurist aids, tools, masks, and talismans as we live through climate change and the steep environmental challenges we face now and ahead.

Author Sophie Strand states:

The mythmaking we are called to do now is probably somewhere closer to composting. We live in a culture that is remarkably good at abstracting itself from waste and off-loading it onto the marginalized communities least responsible for its creation. We cannot simply decide that civilization and patriarchy are toxic and then reject them. Instead, we must take responsibility for our bad stories through the alchemical power of rot. On the compost heap, nothing is exiled. Beliefs and epistemologies that were never designed to touch, combine inappropriately in the moist refuse pile, fermenting into soil that can grow something new to meet the demands of our dire circumstances[1].

Aesthetic Force

William Anders was among the crew of astronauts on Apollo 8 when he snapped a photo from the spacecraft as Earth came up over the horizon with his Hasselblad camera on Christmas Eve in 1968. His photo *Earthrise* changed our relationship with the world forever. In its new year edition, Life magazine printed the photo on a double-page spread alongside a poem by US poet laureate James Dickey.[2]

Harvard art historian and author Dr. Sarah Elizabeth Lewis attributes the power of this photograph, and its ability to cause a seismic cultural shift, to a phrase she coined called "aesthetic force." This image catalyzed the environmental activism movement of the late 60s. And there is a long list of others, including the first time an architectural drawing of a slave ship was published in a newspaper, which illustrated the number of people it was designed to hold versus the number of people it did hold. That image helped to catalyze the abolitionist movement. "Beauty slips in the back door of our rational thought and gets us to see the world differently," writes Dr. Sarah Lewis. She describes aesthetic force, and how through the vehicle of visual culture, it has an impact that is both profound and indelibly lasting. "The words to describe aesthetic force suggest that it leaves us changed – stunned, dazzled, knocked out. It can quicken the pulse, make us gape, even gasp with astonishment. Its importance is its animating trait – not what it is, but what it does to those who behold it in all its forms. Its seeming lightness can make us forget that it has weight, force enough to bring about a self-correction, the acknowledgment of failure at the heart of justice – the moment when we reconcile our past with our intended future selves. Few experiences get us to this place more powerfully, with a tender push past the praetorian-guarded doors of reason and logic, than the emotive power of aesthetic force."[3]

Icelandic–Danish artist Olafur Eliasson, known for sculptured and large-scale installation art employing elemental materials such as light, water, and air temperature to enhance the viewer's experience,[4] said, "I believe that one of the major responsibilities of artists – and the idea that artists have responsibilities may come as a surprise to some – is to help people not only get to know and understand something with their minds but also to feel it emotionally and physically. By doing this, art can mitigate the numbing effect created by the glut of information we are faced with today, and motivate people to turn thinking into doing."[5] It is this elevating, animating, emotive power of it that can motivate the heart and action. Data on climate change, the accelerating pace of species disappearance, and massive ecological destruction is important, necessary, and serves its purpose. But on a mass cultural scale, it can also add to that numbing effect of the glut of information we are inundated with every day. But what motivates the very axis of our beings, creates shifts in perspectives, touches the vitality of the soul, activates the imagination the way art can, creates another avenue to reach one another and revelations of truths, and can create a care-based bond that affects action.

Umwelt

> Bat sonar, though clearly a form of perception, is not similar in its operation to any sense that we possess, and there is no reason to suppose that it is subjectively like anything we can experience or imagine. This appears to create difficulties for the notion of what it is like to be a bat. We must consider whether any method will permit us to extrapolate to the inner life of the bat from our own case, and if not, what alternative methods there may be for understanding the notion…It is difficult to understand what could be meant by the objective character of an experience, apart from the particular point of view from which its subject apprehends it. After all, what would be left of what it was like to be a bat if one removed the viewpoint of the bat?
>
> *Thomas Nagel, 'What Is It Like to Be a Bat?'*[6]

Walking in a city park around sunset, a person might see a bat swoop low overhead, in pursuit of a moth. While the human and the bat are both mammals, warm-blooded, with fur (or hair), it's otherwise difficult to overstate how different they are, and how differently they perceive the world around them. They are mutually aware of one another; the human sees the bat and the bat is perfectly aware of the human's presence. But the human cannot hear the bat, although the bat is constantly calling, emitting a clicking sound, a series of high-pitched chirps. The bat uses these sounds for echolocation or bio sonar – the calls bounce back to the bat's large and elaborately wrinkled ears, which contain elaborate inner-ear structures, with specially reinforced cochlear hair cells that can withstand the amplitude of the bat's calls. The bat's brain has large regions dedicated to interpreting these sound waves, creating a picture of the world around it even in near-total darkness. The sounds can penetrate and perceive the density of objects near the bat. They can pinpoint an object smaller than a mosquito, and determine its rate of speed. Some species of moth have evolved the ability to perceive and recognize the bat's calls and will respond by going into evasive maneuvers or even jamming the bat's echolocation with their own sounds, or by blocking and reflecting the sound waves with their wings.

The bat's calls are so high-pitched that they are above the range of human hearing. This is for the best, as the bat, a Big Brown Bat, in this case, calls at about 110 decibels. This is as loud as a jackhammer or rock concert. It is louder than a person shouting. If several bats are swooping overhead, as is typically the case, the evening sky is a cacophony that would be deafening to us, and in fact, could cause hearing damage if we were able to hear it at all. But the humans below are blissfully unaware. If one of them is walking a dog, the dog is certainly aware of the clangor of the bats, as the dog can hear a considerably higher range than the human. If it is annoyed by these sounds, the dog doesn't show it – the dog is also immersed in its own world of smells, which carry messages from fellow dogs and information about other animals, such as the raccoon that had walked by a couple of hours prior.

The mantis shrimp is a marine crustacean with eyes that are considered to be one of the most complex in the animal kingdom. They have such advanced vision and incredibly colorful bodies that it is suggested their evolution of color vision has taken the same direction as the peacock's train. Their eyes are so remarkable that they can perceive multispectral images and polarized light. The eye itself consists of two flattened hemispheres separated by six parallel rows of specialized ommatidia, which divide the eye into three sections, allowing each eye to have trinocular vision. The ommatidia are arranged in four rows and each row carries sixteen photoreceptor pigments. Twelve of these are for color sensitivity and the rest are for filtering. In comparison, humans have only four visual pigments. Their eyes help them to recognize different types of prey species, many of which are transparent or have shimmering scales. And because they hunt with rapid movements of their claws they need very accurate ranging information and depth perception. One thing this incredible vision reveals most profoundly is that there are a lot of colors that exist in the universe that we cannot even see or experience. These creatures see a different picture.

These spheres of sensory experience create a profoundly different picture of the world, which, as mediated by our senses, has no true objective reality. This perceived environment is known as the *umvelt*, from the German word meaning environment or surroundings. This concept was proposed in 1909 by German biologist and animal behaviorist Jakob Johann von Uexküll, the founder of the field of biosemiotics. It doesn't refer to the 'actual' environment if such a thing can be said to exist, but to the perceived environment, as filtered through any given species or individual's senses, this perceived world can vary and shift in profound ways from one species to another.

How can we learn the unseeable and unhearable things unless we lean on the intelligent systems of all beings around us on this earth who have capabilities so vastly different from ours? And while we can study these capabilities and stand in awe of them, perhaps come to understand them in an academic sense, we can never know what the bat experiences as it echolocates its way through the night with sonar. Nonetheless, simply knowing this vastly different experiential realm exists broadens our own sense of the possible. We cannot know what it's like to be a bat or a mantis shrimp because echolocation and enhanced vision are so alien to us – do the sounds coalesce in the bat's brain into something like an image, or is it more abstract? It's auditory, but is it perceived as sound? How does the mantis shrimp experience the additional color palette that we cannot envision? We can't really know. But the simple act of trying to imagine these umwelten, these fantastically divergent spheres of perception around us, brings us closer to our fellow living beings and pushes the limits of our capacity to imagine.

Visual Mythology

> Those who dream by day are cognizant of many things which escape those who dream only by night.
>
> *— Edgar Allan Poe*[7]

Artists work in the realm of shifting mindsets, perceiving other realities, creating prompts that actively engage the viewer to experience both of these things. Artists' work can roam around within the functions of mythology that can serve as a compass to each generation and are constantly in a process of evolution. Poet Joy Harjo writes "I know I walk in and out of several worlds each day." We, as a society, need our mythologies and our magic. Mythology is encased in a visual story and it can shift the values of culture, from perceived or artificial values to heart-centered ones with integrity. The impact of visual mythology is that it is this living active film playing inside our minds and influencing us consciously and unconsciously, and this directly impacts our actions. The public imagination, activated through the arts, is the primordial space of dreaming together. Art asks us to expand ourselves to be able to not just hold and carry complex coexistence but point out that in fact, this is what we're capable of being made by, not undone by. As poet David Whyte wrote, "The language we have in this world is not large enough for the territory that we've already entered." Part of the artist's role is creating the mythological narratives needed for the context of our times, ones that no longer center only on human concerns and where humans serve as heroes. The creation of mythic webs and networks showing how interconnected we are and the symbiosis of our stories creates internal connections towards a world vastly beyond ourselves.

Metabolizing The Now

A recent article published in the Wall Street Journal titled "Is Looking at Art a Path to Mental Well-Being?" described the growing field of neuroaesthetics, the study of how the brain reacts to different forms of art, and how exposure to beauty in its many forms can affect one's mental health. "Generally, beauty and music or art are very rewarding to the human brain," says Wendy Suzuki, a neuroscientist and professor of neural science at New York University. "It can activate our natural, de-stressing part of our nervous system called the parasympathetic nervous system that slows our heart rate down." The parasympathetic nervous system is the system in our body that is responsible for rest, sleeping, and enjoyment, among other things. When it's activated, you're more likely to be able to think clearly. This effect has also been shown to happen when spending time in nature – a recent study published in the journal Scientific Reports found that observing birds had mental health benefits that could last for up to 8 hours.[8]

The art experience often centers on asking questions, calling on the viewer to look, listen, and be curious about what's happening, where, what's driving it, etc. By going through this process, a viewer might see a new way to think about something, face something and be less afraid of something. Robin Wall Kimmerer writes about the importance of restorative practices, stating, "We need acts of restoration, not only for polluted waters and degraded lands but also for our relationship to the world. We need to restore honor to the way we live so that when we walk through the world we don't have to avert our eyes with shame, so that we can hold our heads up high and receive the respectful acknowledgment of the rest of the earth's beings."

Situating the art experience as a place of stopping, questioning, facing, and reframing, what is also created is a space for a lack of judgment, of oneself and of others. Removing this helps with clearer thinking, dialogue, and idea generation. It is a well-known adage that when one faces their fears, it helps to diminish them. And existing, being incrementally exposed to and in dialogue about what is happening to the earth, or experiencing the world from a non-human perspective, creates space for us to process. Watching environmental, economic, and resource disasters that we have seen this past decade, an inundation of videos and images we see on the news and social media outlets daily, one can sense where these deep fears of our time come from in the cultural imagination. They need to be counterbalanced with experiences for us to process them, which is an area art can occupy. Art also has a place in helping us tend to the soul wound and grief of witnessing climate change and what is happening in our environment. Art serves a purpose for us as transformational so that we can then tangibly have a better grasp of understanding. And perhaps "facing" within the imaginative realm aids the strengthening of our muscles to face these challenges in reality.

Eliasson also stated, "Art also encourages us to cherish intuition, uncertainty, and creativity and to search constantly for new ideas; artists aim to break rules and find unorthodox ways of approaching contemporary issues."[9] This is also speaking to the reorientation and connection transformative art experiences can have, something that lands in the body, mind, heart, soul, uniting and activating ALL of the complex wisdom systems that reside in us, not just raising the rational mind as the one at the top of a hierarchical framework. In elevating intuition, uncertainty, and creativity, ideas can expand, ideas on how to face the vast array of ecological challenges in our world right now and ideas on how we can adapt ourselves. Art ignites the power of imagination and vision, and within a society hijacked by the abstracting, rationalizing, and, perhaps most dangerously, controlling ego.

I employ intuition a great deal in the studio. It acts as a guide and many more surprising relationships and combinations of things arise when I do. It is leaning into this space of uncertainty that opens more doors to creative innovation (Figure 15.2).

FIGURE 15.2 Vessel, 2021, Sarah Walko

Remember

> How monotonous our speaking becomes when we speak only to ourselves! And how insulting to the other beings – to foraging black bears and twisted old cypresses – that no longer sense us talking to them, but only about them, as though they were not present in our world… Small wonder that rivers and forests no longer compel our focus or our fierce devotion. For we talk about such entities only behind their backs, as though they were not participants in our lives. Yet if we no longer call out to the moon slipping between the clouds, or whisper to the spider setting the silken struts of her web, well, then the numerous powers of this world will no longer address us – and if they still try, we will not likely hear them[10].
> — *David Abram, Becoming Animal: An Earthly Cosmology*

Part of the artist's role and responsibility now, with these threats to our very existence, the existence of other species, and the existence of earth itself, is concerning the beingness of all. Art aids us in processing and metabolizing current situations. If we cannot fully face, take in, and digest our current challenges, we cannot fully and effectively address them. Art contains an inherent healing property but it has the unique ability to go past this into elevation, celebration, and correction and this can extend into restoration, regeneration, and reciprocal relationship. It can function as a liminal space preparing us between states for new beginnings and new beingness.

Notes

1. Sophie Strand, "Rewilding Mythology," *Atmos*, September 15, 2022, https://atmos.earth/rewilding-mythology-mythmaking-climate-collapse/.
2. Joe Moran, "Earthrise: The Story behind Our Planet's Most Famous Photo," *The Guardian*, December 22, 2018, https://www.theguardian.com/artanddesign/2018/dec/22/behold-blue-plant-photograph-earthrise.
3. Sarah Elizabeth Lewis, *The Rise: Creativity, the Gift of Failure, and the Search for Mastery* (New York: Simon & Schuster, 2014), 93–94.
4. Wikipedia, "Olafur Eliasson," accessed December 18, 2022, https://en.wikipedia.org/wiki/Olafur_Eliasson.
5. Olafur Eliasson, "Why Art Has the Power to Change the World," *World Economic Forum*, January 18, 2016. https://www.weforum.org/agenda/2016/01/why-art-has-the-power-to-change-the-world/.
6. Thomas Nagel, "What Is It Like to Be a Bat?," *The Philosophical Review* 83, no. 4 (October 1974): 435–450.
7. Edgar Allan Poe, *The Fall of the House of Usher and Other Tales* (New York: Penguin Random House, 2006).
8. Ryan Hammoud et al., "Smartphone-Based Ecological Momentary Assessment Reveals Mental Health Benefits of Birdlife," *Scientific Reports* 12, no. 1 (2022), https://doi.org/10.1038/s41598-022-20207-6.
9. Eliasson, "Why Art Has the Power to Change the World."
10. David Abram, *Becoming Animal: An Earthly Cosmology* (New York: Vintage, 2011).

16
ADRIAN PARR ZARETSKY IN CONVERSATION WITH CARLA BENGTSON

Adrian Parr Zaretsky and Carla Bengtson

AP: Your work is deeply collaborative, in particular I am thinking of your unique approach that involves working with members of the scientific community. What does this process look like? And what prompted you to move beyond the realm of aesthetics and into the sphere of scientific discovery as a mode of art making?

CB: My transpecies projects emerge and find form through a speculative conversation between myself, my human and other-than-human collaborators and the sensory forms and material possibilities that suggest points of intersection between human and other-than-human nature(s)/culture(s). The work typically begins when I come across something surprising about another species. This can happen spontaneously through something I happen to come across, or in conversation with a scientist, or because I've been researching or noticing a species in an area I'm visiting. This is followed by a period of extensive research. I often reach out to a scientist that is studying the species I'm interested in, which can result in a series of exploratory conversations. Once trust is established, the relationship might lead to a collaboration. The conversations with scientists are critical because scientific research papers usually have a relatively narrow focus. More informal conversations can lead to a broader understanding of the other-than-human animal's evolutionary ecology, temporo-spatial behaviors, sensory modalities, cognitive capacities, and social dynamics. It can also lead to my posing some speculative questions that we might address together. Lately I've been asking, "What questions do you think the animal itself would want you to ask of it? Or it of us?"

These scientific insights and speculative questions lend a level of specificity to my creative decisions, and result in a number of creative propositions to try out in the field. My field experiments are purely speculative, more in the nature of play than of controlled experimentation. Usually it means some kind of intervention into the other animal's umwelt, or lifeworld. This again is where working with scientists is valuable, as they can help me to understand how the animal-other might respond, what it might mean to them, and help to ensure that I do no harm. What I'm looking for is some kind of bridge to, or response from, the other-than-human animal. The field work is very slow, often physically challenging, and requires a lot of patience, as any field scientist will tell you. But

DOI: 10.4324/9781003403494-19

sometimes the response is quite astonishing. And even their refusal to perform can become the focus of a piece.

Once I get a handle on what interests me, what the animal has to say to me, I may initiate collaborations with dancers, musicians, perfumers, philosophers, or anthropologists. These further collaborations open up a range of embodied pathways that parallel or intersect with the lifeworlds of other creatures, as well as disciplinary perspectives that help us to reflect on our own positionality. I think of this triangulated relationship between scientists, species, and artists as a third space of inquiry where a variety of disciplinary methodologies, sense-related creative forms, and nonhuman perspectives are in dialogue – a space in which previously unthought and unfelt sensations, experiences, and perspectives can be returned from nonhuman nature/culture(s) to human nature/culture(s).

In terms of my relationship to aesthetics and scientific inquiry, I would say that I haven't moved beyond aesthetics, I've simply reframed aesthetics to include the aesthetic experience of other creatures. This is, at heart, a speculative proposition since we can never truly know. But the *attempt* to know is in itself a gesture towards inclusivity. The gesture expands our focus and creates a sense of relation and kinship that makes room for difference.

I also find that science and art complement one another. Science, like art, is a speculative form of inquiry. But its means and its ends differ from those of art. For me, science offers insights into the sensory modalities, communication systems, and social relations of other creatures that I might miss through first-hand observation or be occluded by my own human-centric perspectives. At the same time, science lacks a methodology that can address how we feel towards other creatures, or that can embrace their mystery and alterity. In a way, science is a call to knowing more, and art is a radical allowance of knowing less. Of allowing ourselves to exist in the liminality of wonder. When the two are combined, we are able to exist in a space of curiosity that suspends us between knowing and not knowing. The knowing can act as a means of access, and the not-knowing can make room for the animal-other's alterity and their agency. Their right to exist in a world entirely apart from our own.

AP: Your work combines a number of elements such as everyday materials, bodily fluids, along with animal and plant life, producing imaginative encounters that engage the myriad ways human beings experience desire and sensuality. How do you approach integrating and combining very different mediums?

CB: I'm always following the animal's lead in terms of the choice of media. That might mean bringing in scent, sound, taste, movement, or touch, depending on the other-than-human animal's primary sensory modalities and communication strategies. What I'm looking for are parallels, or points of intersection, between how human and nonhuman animals communicate and experience sensuality. It's not anthropomorphism. Almost the opposite. I'm equally interested in making familiar as in making strange. In stepping away from ourselves for a moment in order to enter into a speculative relationship with another creature. Meeting other species in all their strangeness and wonder also means looking back at ourselves, seeing ourselves from the eyes of another creature – if they have eyes – in all our strangeness. Bringing everyday materials into dialogue with organic materials reminds us of our hybrid nature, of our condition of being always between and within our bodies and our cultures.

The *Euglossa sp*.orchid bee perfume bottle (Figure 16.1). in the exhibition is an example of this merging of generative relationships that point to different realms of influence.

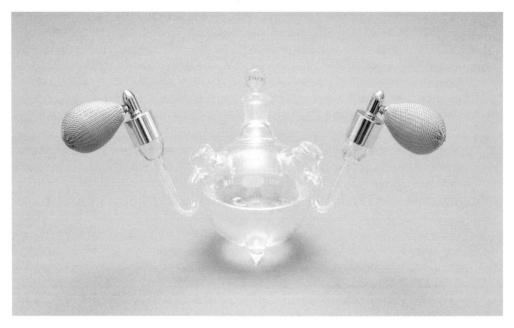

FIGURE 16.1   Carla Bengtson, *Euglossa sp.* (2016), handblown bottle containing a reproduction of the perfume male orchid bees use to attract a mate; glass, metal, cloth, natural scent compounds. Photograph: Jon Bagby

        The bottle's form and function point to how the male orchid bees disperse their perfumes for females by fanning their wings, which atomizes the scents. But the bottle is a blend of science lab and perfume industry elements, so it points in all three directions at once.

AP: Your work with odor as an artistic material, introducing new sensorial experiences into the field of art appreciation and reception. Smell can be a deeply sensuous experience as much as it can be unpleasant and even offensive, depending on the scent. Do you place these seemingly contradictory experiences in tension with one another as part of the work, or do you see your practice as exploring the power relations that structure our experiences of pleasure and repulsion?

CB: What attracts me to scent as a medium is the extent to which it bypasses language. It is our most primal sense, yet it touches memory and culture. What also attracts me to scent as a medium is that we physically take scent molecules into our bodies with our breath. Which means that, in a sense, we become it. So it really is a kind of mutual becoming.

        An example of a project where I've used scent, along with taste, sound, and gesture to create a human/animal becoming is *Every Word Was Once an Animal*. The exhibition was inspired by *Sceloporus* lizards, also called fence lizards or blue-bellies. Fence lizards use gesture, along with scent, to communicate. Their gestural language uses patterns of head-bob and push-up movements that, like human language, are socially learned, make use of syntax, and exhibit a range of regional dialects and subtle individual embellishments. My collaborator, Dr. Emilia Martins, uses lizard robots programmed with lizard language to decode their movement patterns. And she has also taught me lizard language! To bob my finger in imitation of their species-specific movement patterns to evoke a species-recognition response. I've made a hiker's bandana with a lizard language decoder so others too can speak the embodied language of another creature.

But, back to scent. The scientific name for fence lizards, *Sceloporus,* means scented thighs. Along with gesture, fence lizards secrete pheromones from their thighs that are used to claim territory, express their identity, and attract a mate. Dr. Martins has analyzed their primary scent compound and discovered that they include jasmonates, a common ingredient in human perfumes, as well as pyrazine, which give white wine its distinctive grassy notes, and are found in citrus, coffee, and chocolate. For the exhibition I created a perfume based on the lizard scent compounds. It's a fresh, green, floral perfume that responds to the wearer's own skin chemistry to create a soft, sheer, more-than-human presence. During public events exhibition goers were invited to sample the scent while sipping white wine and tasting the same scent compounds in jasmine, grapefruit, and coffee-flavored truffles. Dancers entered the gallery intermittently to perform an invented gestural language, while five singers performed vocal responses to the head, heart, and base notes of the scents to create a scent/sound/taste/movement synesthesia. Nonhuman animals, unlike humans, most likely experience all their senses at once, rather than focusing on one at a time as we have learned to do. Although in the Western world we tend to suppress our sense of smell, our olfactory response has been shaped by evolution to be a survival tool, producing immediate reactions of attraction or repulsion. The shock of finding pleasure in the secretions of a lizard, actually taking some of their scent compounds into our own bodies, radically destabilizes our understanding of aesthetic experience.

Besides attraction and repulsion, there are other ways in which people and other-than-human animals use scent. I'm currently working on a project with prairie dogs, known to have a highly developed sonic and gestural communication system that approaches our human definition of language. Besides sound and gesture, prairie dogs use scent to create important social bonds. Their altruistic alliances are reinforced by sharing one an other's scent during extended greeting behaviors that resemble kissing, and by sampling musk-like scents from their anal glands. In addition to using scent to establish social ties, some species of prairie dogs are known to anoint their bodies with rattlesnake scent as a means of concealing their odor from their rattlesnake predators. They do this by chewing on molted rattlesnake skins (which smell like cucumber), then licking their fur, which masks their scent while they are underground. I'm working on a rattlesnake camouflage perfume for humans. It will have animalic base notes of musk and fur, grassy prairie middle notes, and top notes of rattlesnake. I envision a social sharing of the scent to promote animal-centric alliances among humans, who will be asked to volunteer or donate to organizations that protect endangered and threatened species of prairie dogs.

As we become less and less reliant on personal experience, and more and more on mediated content, most people don't have an opportunity to form a personal relationship to the problems of climate change and species extinction. As a result, these problems seem to exist outside ourselves, or in someone else's hands, and we don't take action. Experiencing animal-related scents, even taking them into our bodies, offers an opportunity to perceive and relate to species that are otherwise invisible.

AP: If you look back on your studio practice over the years what experiences prompted you to expand your art practice beyond the realm of human creation into the realm of other-than-human species? How might the experiences of other-than-human species move beyond being the subject of art and into the realm of co-creation?

CB: I was an abstract painter for many years. Paint has this wonderful capacity to be both other and self. To be intoxicatingly actual in its sensuality, and cultural in how it carries its history on its back. For me, it was the meeting place between my senses and my intellect. An exquisitely sensitive arena that I attempted to rethink from feminist, materialist, and phenomenological perspectives. But over time, I started to look more towards my experiences in nonhuman nature to see if it could help me to unlock the hold that culture has on culture. To see if it could teach me to see it, to step aside from my own self-reflection, to understand its necessity and its alterity.

Around this time I met my husband, Peter Wetherwax, a pollination biologist and frequent collaborator, and together we spent the first of what became thirteen residencies at a Biodiversity Research Station in the Amazon. Experiencing that overwhelming mass of buzzing, whirring, growling, singing beings speaking in animal tongues and senses caused me to shift my focus from what I see, to what sees me. And from what I think, to what thinks me. Everywhere I looked there was a creature claiming life, vying for a foothold, even down to colonizing my skin. All this came to a head when I spent a night alone in a tree watching the glow of fire on the horizon from an illicit oil well. When I awoke in the morning, a group of woolly monkeys had moved into the tree with me and were peeking at me from behind branches with questioning looks. They seemed to be trying to understand what I was doing there. I had to ask myself the same question. At that moment, abstract painting was no longer relevant.

The second part of your question, "How might the experiences of other-than-human species move beyond being the subject of art and into the realm of co-creation?" is one that I've been asking myself ever since. The first works I did following my encounter with the woolly monkeys were with leaf cutter ants, who produced ant track drawings for me that were curiously reminiscent of modernist painting, but which meant nothing to the ants themselves. So I made paintings based on their ant track drawings, blew them up, and returned them to the site. They promptly adopted the smooth surface of the print as a sort of super-highway, while others cut and harvested the print to feed their fungus farms (resident ant researchers provided guidance on how to do this without harm to the colony).

That project was followed by "Snake Detection Theory," a tongue-in-cheek attempt to help monkeys evolve by teaching them to point at snakes. Which, according to primatologist Lynn Isbell's snake detection theory, was a form of directed social attention that nudged our hominin ancestors towards language.

In both pieces, the ants and monkeys were co-creators in the performance of meaning. Theirs to me, and mine to them. The orchid bee perfume in the exhibition is another example of a material co-creation that resulted in an aesthetic experience for both. The bees offered me the inspiration for the perfume, and in a reciprocal gesture I returned my perfume to them. The bees sampled and collected my human perfume and added it to their own scent blends, which were used to attract mates. So from the very beginning I have tried to move beyond subjectivizing the animal-other to learn what the animal itself might want, what it has to say to me.

AP: In my own work I am interested in how transpecies design might present us with an opportunity to move beyond an anthropocentric approach to creative production. In particular I put forward three principles that underpin this kind of work – regeneration, restoration, and reconciliation. In many respects your work uniquely blends all three principles in the ways that you massage emotion, activate memory, and sensitively engage difference.

Can you describe how you envisage your work engaging with these three principles of transpecies design?

CB: My work is very much aligned with the principles of regeneration, restoration, and reconciliation, and with your strategy of assuming positions of equity, friendship, and generosity as starting points for change. It is meant to resituate us conceptually in relation to other creatures. To recognize kinship through our shared corporeal modes of being, and to shine a light on animal voices and nations. My hope is that it can remind us that the world is continually being made and remade, by us and by our fellow creatures. And with this realization, we begin to understand that it can be remade to be more just, more life-giving, less selfish.

AP: What art, music, and novels have inspired you and how have these shaped your practice?

CB: This will seem from way out of left field but the art that has inspired me the most over the years is the brief window of work by Russian Avant-garde artists such as Malevich, Popova, and the constructivists that happened in the decade or so following the Russian revolution. I think that their project has been dismissed as being merely utopian, but what I have taken from it is a desire to make the viewer active, rather than passive, and the theory of "making strange." Also the conviction that aesthetics can either passively reinscribe the dominant ideology, or actively search for new terms of possibility, even at the level of formal relations. For me, art is an ongoing practice in active, open-ended engagement for both the maker and the viewer. This was the proposition that I pursued as a painter, and the proposition that I continue to pursue. It's just that my attention has shifted from what and how I think, to what thinks me.

Philip Glass's music has also been a huge inspiration. His work inhabits that place the playwright Bertolt Brecht described as "thinking feelings" and "feeling thoughtfully." That's where I like to live. Somewhere between complete sensory immersion and intellectual awareness that pushes up against the boundaries of what can be thought, felt, or said.

Octavia Butler's idea of "primitive hypertext" resonates with me. It evokes a state of mind that is at once attentive and meandering, non-linear and associative, and full of possibility. It seeks out the generative relationships between wide-ranging ideas, objects, and experiences and holds them in open-ended suspension. That's what I'm trying to do when I bring human and nonhuman forms of cognition, language, sociality, materials, and aesthetics into play with one another. And of course, I like her other-than-animal-like superhumans.

17
PIÑON PASSAGE

Nina Elder

It was mid afternoon on the winter solstice. I had just returned from saying goodbye to my dying father. I walked into the hills. My lungs were heavy. I was recovering from a sickness that I had mistaken as grief but revealed itself as covid. I moved slowly, with the tentative ambition of the newly healthy, up steep hills that are made of lava rocks and dust. The low angle of the winter desert sun scraped light and shadow with elongated gestures through the wind rumpled grasses.

FIGURE 17.1 Nina Elder, Piñon Skeleton, photograph, 2022. The desiccated form of a two-needle piñon is silhouetted against the bright desert sky.

Linear blue shadows striped amid symphonies of yellow. Buttercream, ochre, blond, copper, vanilla, lustering lemon, eggshell, cadmium, mango, sulfur.

In this needling sideways spotlight, sap seeps on the piñon shone bright. Petrified light nestled into scaly bark. Amber scabs sealing off the inner tree from the marauding outside.

For 400 or 500 or 1000 years, this yellow blood had been coursing from root to branch tip, transporting sugar and proteins and lifeforce and magic. Now the piñon are hemorrhaging sap. Scars, holes the size of hungry tiny beetles, gouges made by chainsaws, dismemberments caused by wind or age, each wound evidenced in oversized golden keloids of arboreal blood.

Stilled showers of falling amber. Nodules. Blobs. Drips. Oozes. Accretions. Tumors. Incrustations. Bejewelments. Tree tears.

I began gathering every burble of sap that I found on the dead trees. Prying and plucking with my right hand, piling into my left hand. What good is a scab on a dead body? What flow is there left to staunch? By the time I reached the overlook, I had an accumulation in hand about the size of a human heart. On the shortest day of the year, I was holding centuries of distilled sunlight.

Climbing uphill, my body heat had begun to affect the jumble of sap. Remembering its once viscous vitality, it filled the space between my fingers and the lifelines of my palm. It was gluey and strange. Perfume clouded the air around me. Pungent, sweet, acrid, incensed, heady, holy. The secret smell of the dark inside of a piñon tree.

Death, in my warm hands. Something solid became soft, releasing an enduring sweetness, a terrible stickiness.

The two-needle piñon (Pinus edulis) is so common in the American Southwest that we all spell it differently (Figure 17.1). Pinyon. Pinion. Scrub pine. Nut pine. T'o in Tewa. Be'ek'id Baa Ahoodzání in Navajo. I choose piñon because the enye makes sense in my mouth – it is the shape of my tongue as I roll through the middle of the word. It is the cirrus clouds over humpy hills that feel familiar. It is the brevity and lyricism of New Mexico.

The juniper-piñon ecosystem covers over 10 million acres and ranges from Texas to California. The compact trees have long lifespans, many of them older than the colonial names and geopolitical boundaries that surround them. The piñon and juniper combine to create iconic dark green polka dots across western mesas and red rock country. Their fruit, the soft sweet pine nut, are integral to the diet of thousands of species of birds and animals. Piñon sap is a healing salve, its wood is fuel, its structure and sinew are construction materials. Piñon are ubiquitous, an epitome of keystone species.

From a distance, piñon appear small and shrubby and identical. But walking among them, one realizes they often reach twenty feet or more in height, a maze of twisting trunks and branches. They are densely packed with life and whimsy, with braided lumpy reddish gray bark. Needle bundles grow haphazardly from all surfaces, not just the twiggy tips. Needles emerge and shed in irregular 4-6-year cycles, evergreen.

The two needles of the piñon are actually two halves of a leaf cylinder. When they are soft and young and closer to chartreuse, this bundle appears to be a single, straight, pointed, two-inch-long barb. The fascicle from which the needles sprout is a hard granule encased within tissuey wisps. As the needles mature, they darken and lean ever so slightly apart, a glimpse of sky slipping between the two halves. Throughout their time together, the bifurcated needles spread apart and draw close and spread again. In stress, they harden and curve away from each other.

In the predawn dark of December 26, my sister came to wake me up. She said our stepmother had texted, asking us to call. Our dad had promised her that he would stay alive through Christmas; he was a man who kept deadlines. Somehow in the groggy gray cold, I knew to say, "not yet." I was not delaying the grief – that had already grown and diminished through the 20-year evanescence of my dad. I wanted my sister and I to be aligned, attuned, energetically one. Before we made that inevitable phone call, she laid down next to me for several long quiet minutes. Her shoulder blades a breath away from my sternum.

When a piñon turns orange, it happens quickly. There is no charmed unfurling of death, no whisper slow funeral song. I have turned my back on a green tree, and no motion in my periphery turns me towards its dying. In a glance, it is orange.

Piñon need the cold. In a deep frozen stillness, their sap slows, and the beetles who dizzy dive into their nectar die. New Mexico winters have become a tedium of sunshiney sameness, sixty degreeness, skin wrinkling dryness. Beetles feast.

Piñon need fire. It takes over 150 summers for the trees to begin bearing fruit, so occasional fires allow them to mature with less competition or strain. Their thick bark can singe without incinerating, leaving the lifeforce of the piñon unscathed. Humans fear fire. We forget it was our gift from Prometheus, our godly charge. We have tamed fire into torpor. Combustion in engines rather than as components of healthy ecosystems leaves piñon struggling.

Orange needles are a temporary cloak. Winter winds defoliate the dead. I reach out and touch a dangling set of brittle twins and they spin to the yellow grasses at my feet. They fall together.

For a few seasons, the dead tree will hold its skeletal form, drying, darkening, an always silhouette in this bright place. Then branches will fall, packrats and birds will scuttle away with the twigs, and desiccated roots will wend skywards from the sandy soils. The stumps tip and bow to the elements. Roots silver in starshine, windharrowed to dust.

My sister and I are two halves, sharing a root, just a gasp between us. We lived together for most of our adult lives, hopping from Colorado to California to New Mexico. Now, I travel nearly full time for research and work. She allows me to do laundry at her house, sleep in her art studio, fill up endless water jugs from her hose spigot. One of my homes will always be with my sister. One of my homes is silence and sky. One of my homes is in the newness and discovery of travel.

My sister often does not raise her arms to hug me as I wrap around her self-contained torso. My sister is privately rigid, a stiffness that comes from a deep rusting place between us. My sister makes people guess which of us is older. I am younger, which few surmise. I am taller with haggard hard edges from hard years, hard laughter, and hard sunshine.

My sister loves me, but she often does not like me. She would push me much further away if she could. I would fully embody independence if I could. But we share a root, we are two halves. We are bound, sometimes supple, often destructively brittle. Distance and demise can feel like synonyms.

The night is loud and windy. Exhalations freeze on the windows, dark truths from inside my body expressed in white whorls. Piñon burns in the woodstove. Inside the black metal, small explosions of boiling sap snap pop fizz hiss. The flames release hundreds of summers' worth of heat into my drafty metal trailer. The smoke is pulled out into starlight, a horizontal stripe of incense and carbon. No one will smell the musky scent, no one will cross the perfume perimeter of my safe warm home. It wings into the peopleless night, a sweet story told, unheard.

I am the punctuation point at the end of two families. Of my generation, I am the youngest. Only one cousin procreated and made two new humans, a sideways branch sticking out of an otherwise thoroughly pruned family tree. In me, there are bloodlines tangling without tangenting, DNA mixing and mashing without ever expressing. I imagine a large triangle of humanity balanced on my head. In an hourglass of heredity, a pile of my childless cousins and sister and me block off any passage. No more red hair, anxiety disorders, breast cancer, boisterous laughter, and boundless curiosity. A shrinking choir of strong alto singing voices.

Childless women are third, after men and mothers. Our biological purpose and privilege of being is questionable.

Yet,

I rejoice in my singularity. I have never wanted children.

In our childhood, my sister and I knew no women who were unmarried, without children. We were only presented with heterosexual, proliferating, partnered parents as options for role models.

We choose the unknown, the unmodeled.

Science predicts that 80% of piñon trees will die by 2060. In warmer parts of New Mexico, there will be a complete eradication of the species. Grassland will succeed the piñon forest as the dominant ecosystem.

I felt relief and bittersweet recognition when I learned about solastalgia. A newly named psychological condition, solastalgia is the distress felt for impending changes in the homeplace, a *pre*-traumatic stress condition. It is homesickness felt before leaving home. It is a premonition of grief. Solastalgia is knowing that I can't stand on the bank of this river forever because floods will come and erosion will yank at my feet.

For more than a handful of years, I have traveled. I teach, I talk, I guide, I learn. I make drawings. But a demanding reality showed up like a cactus spine in my sock. I realized that I was fleeing from solastalgia, even after I had written articles and made exhibitions around the concept. If I had no home, I could not feel the grief of my home place changing. If I had no roots, they could not go dry. I had preemptively severed myself from future suffering. I live in contradictions. I hold audiences' attention by acknowledging climate grief and sharing creative responses, yet I cross the country every month, guzzling fossil fuels and gas station coffee. I can discuss complex ecological issues, yet I stammer when someone asks me where I live. If utopia is no-place, I was living in it. Not quite ignorant bliss, but I was moving with gail force gusto through the hardest of topics.

Piñon pollen can travel up to 2000 miles, and moves on winds up to a mile above the surface of the earth. The pollen's range often extends far beyond its own productive viability.

My sister and I both own chainsaws. We similarly enjoy hard work and self sufficiency. We both are cautious. I waited until she could visit to cut a dead piñon on my land.

The first remarkable thing was the permeating presence of noise – I often live in unbroken silence for days. The second remarkable thing is how large the pile became, even though we sorted our cut sections by size and dryness. A tree is a well organized and tidy accumulation of matter. The third remarkable thing was how quickly the void went away. For less than one day, there was a place where a tree used to be, a surprising new space for the sky to dip down and touch the grass. Then that place became less of a place, with nothing to hold my eye or distract the wind. How easy it is to forget a tree in a land of dying trees.

When I am not teaching and traveling and talking, I choose to live with as little human contact as possible. I backpack into canyons, roam around on glaciers, circumnavigate mountains, walk into clouds. I seek where the roaring silence of nature muffles the cacophony of society. I try a wilder way, when the veil between human and nature is thin, and I can lose parts of myself. I have felt thirst nearly kill me. I once woke from the permanent blue shiverless cold bright place. I know what it is to have my body break. I have groaned at my human ineptitude and gripped at my lifestrings while saguaros and mice and ice worms and orchids and trees live and die and live and die and live and die around me.

I have understood, with my body, what ecology is. It is not some groovy psuedointellectual flimflam belief, but an entirely real network of life. Cause and effect and coexistence are undeniable in ecosystems made of melting glacial ice or brutalizing Mojave sand shine. There is always something living in the cracks, flourishing in the shade, slurping up the sunshine, growing in the waste, existing in the excrement. Things thrive together. Nowhere is sterile. In extreme places, the fierceness of coexistence is palpable.

Mabel was the most exquisite dog I have known. Not a good dog, but an exquisite dog. She was my sister's dog, but I loved her. In her prime, she was glistening muscle, leaping, black and white, an Orca of a dog, proud and loud, a dog that held space. Her fur was velvet, her eyes luminous, her strength legendary. She used to jump the fence and go two doors down to keep an old lady company. The woman was in late-stage dementia and her son was always startled to find a gigantic pitbull sleeping at her feet. We have no idea how Mabel found the old woman, but she went where she was needed. A fierce friend. She was always wounded from chasing coyotes, bashing through barbed wire, fighting other dogs, and noshing porcupines. She was impervious to pain. Torn skin, bloodshot eyeball, broken toenail, skunk stunk, so much messy life. Mabel the anodyne. Mabel the scary and strong. Mabel the soft. Mabel the unconditionally loved.

The piñons' main predator is the *ips confusus*, the highly evolved bark beetle that is known as an engraver beetle. Making the most of an environmental symphony of stress – drought, suppressed wildfires, increased winter temperatures – they attack and thrive within the structure of the tree. In an ecosystem where things often desiccate and disappear in the wind, a stable *ips confusus* population helps break down tree skeletons into soil. They used to inhabit only the dead and dying, but now they are on a moribund munching frenzy. A male *ips* will bore into the bark of the piñon, choosing trees that are thirsty, running low on sap. His pheromones invite an orgy of procreation, with many female *ips* showing up in the nuptial chamber, breeding, and burrowing out larval nests. These tunnels, or galleries, radiate out from a central line, forming tree-shaped drawings underneath the bark of the tree. Due to the lack of cold, there are now often 4-5 egg cycles per year instead of one or two. The success of *ips* broods in current environmental conditions is astounding, an epidemic. The mazes of tunnels under the bark eventually form a girdle, a stripe of dying material around the circumference of the tree. As spring comes, and the sun warms the mesas and mountains, sap begins to flow upward, delivering nutrients to the needles and new cones. But a rigid belt of inflexible scar tissue holds tight, strangling, squeezing, suffocating. Desperate blobs of amber push forth, extruded below the stricture, every pore and hole expressing some pressure. And the needles turn orange.

The word "ecology" presents the act of living together. In Latin *oeco* means "coming together," which derives from Ancient Greek οἶκος, which translates into "household." When I imagine all that comes together to make a home – shelter, nourishment, rest, safety, a place to love and create and share – I know how to recognize this for other species, but not for myself.

I don't need a refrigerator, a coffee table, a roof, or a bed to feel at home. I have exhausted the easy usefulness of things like light switches, men, garage doors, flush toilets, partnership, thermostats, and junk drawers. I know what I do not need, but am still figuring out what I require.

When grief caught up with me, it was both singular and multiple. It was uncontained and wild, like mercury spilled on an uneven floor. It shone bright, illuminating the low spots. After every artist talk, someone asks me "how do you maintain hope?" and I blurt out "I don't. My heart is broken. But I believe in beauty."

I recoiled from conversations about solutions. At conferences and symposia, I was surrounded by people who were raising millions of dollars to bulldoze and bloviate their hypothetical way towards elusive climate crisis remedies. Sleekly disguised roughshod zigzag ego trip goose chases through crumbling ecosystems.

Meanwhile, the news was full of stories about thousands of people dying of covid, alone, unreached by care or connection. No goodbyes. Just death rattles. Meanwhile, the medical world was racing to find a cure and the healthy were constructing cliques and conspiracies about masks and mandates. The discrepancy between the individual experience and the social experiment was stunning. Spun aswim in depressing mega data, I wondered how it would feel to look someone in the eye and say, "this is terrible. This is not how the world should be, but this is how the world is. You are going to die. I can't fix anything. I am so incredibly sorry." And then stay. Until the end.

Can the lived experience of death inform how we inhabit a future that will be defined by loss?

My sister occasionally gives me lists of rules that I must abide by. Topics to not be discussed, entire brackets of emotions that must not be shared, ways of being that she will not endure in our sisterhood. She can not be with me as I process and prune certain pains. She can't resist picking at my thickening amber flows. She finds herself sticky from attempting to enter my seeping scabs. I am not a tidy person. My body weeps. All rules contain an "or else;" my sister's resonate with an implied future silence. A punishing peace. I lean out, farther.

I looked and looked for a home. I stayed up too late scrolling through real estate listings. I picked up Thrifty Nickels and every other rural community newspaper. I tied notes to falling down fences. I gave my phone number to ranchers. I drove down long dirt roads. I called strangers from online advertisements and roadside signs and gas station notice boards.

New Mexico? Of course.
Acres? Yes.
House? Probably not.

Exposure to the warmth of southern sunshine? Definitely yes.
Cheap? YES.
Neighbors? NO.
Utilities? I don't care.
View? Yes.
Rural? Yes, please.
Dirt road? Fine.
Commute time? The more, the better.
Trees? Oh. Oh, no. I can't handle that question. My heart. My heart. I can't do this.

For Sale by Owner: The real estate listing was bad. Really bad. The two photos were blurry, one was upside down. It had been listed for over a thousand days. The only descriptors were, "No water. No electric. No building site. Cash only."

I went to look at the land as Mabel was dying. Tomorrow, I would say goodbye. Today, I needed to give my heart something with potential and give my sister time to say goodbye. Mabel was done living – she was bloated with blood, leaking fluids from strange sores, losing herself in the opaque haze of sickness.

I drove back to Albuquerque, anxious to be with Mabel and my sister. The next hours were quiet, mainly communicated through touch. Black nose with the texture of asphalt. Goodbye. White whiskers quiver. Goodbye. Massive paws, simultaneously tender and tough. Goodbye. Scarred sweet belly. Goodbye. Deep eyes, now dark pools stagnating. Goodbye. Flanks of flaccid muscle. Goodbye. Once terrific teeth. Goodbye. Soft breath. Goodbye. Softening. Goodbye. Soft slipping into not life. Goodbye. A good and sad stillness. Goodbye.

Grief touches all emotions and coexists in the strangest combinations. It startles. It just shows up. Pets are an especially good proving ground for solastalgia. We know we will outlive them, yet we love them. We love them. Mabel's was my first big death. I was so lucky to be halfway through life and just learning how to grieve. And the biggest surprise was that it felt like home. That ancient Greek οἶκος, that household. The place that can hold grief can hold me.

The day after I turned 40 years old, I was digging the hole for my outhouse, sweat burning stripes through the dust on my face, muscles engaged in a mountain. I suddenly owned nearly 12 acres of high desert. A place with no visible human infrastructure and an unobstructed view of mountains and plains and sky. And piñon. Unbound acres of piñon (Figure 17.2). Beautiful, dying piñon.

I found my home among the dying trees, with the dying trees. Because of their dying, I could finally feel my homing. I was slow to name this, but I will do it. I am aligning the final years of my life with the final years of piñon life. I am twinning my roots with theirs. My hands are sticky with warm piñon sap. I will keep traveling and teaching, but when I get home, I will watch them become orange. I will witness as the piñon change from tree to wood. I will touch piñon with my hands every day. I will split their dead flesh with a chainsaw and an ax. Their hundreds of summers will burn spark bright and keep me warm. Their perfumed prayers will linger in my hair and hide in the hood of my sweatshirt. I will pair my small extinction with theirs. The stoppered hourglass. The children that I never wanted to have, the grasslands that will come, the sisters that fall together.

						2022 Inventory
Stump	Green	Orange	Orange	Green	Skeleton	
Green	Skeleton	Green	Green	Green	Green	
Green	Green	Green	Green	Green	Green	
Orange	Green	Green	Green	Green	Skeleton	
Green	Skeleton	Green	Green	Green	Green	Stump
Orange	Orange	Orange	Stump	Green	Green	Skeleton
Green	Stump	Green	Green	Orange	Orange	Skeleton
Green	Stump	Green	Skeleton	Orange	Skeleton	Green
Orange	Stump	Green	Green	Skeleton	Green	Green
Green	Skeleton	Skeleton	Green	Green	Orange	Orange
Green	Orange	Green	Skeleton	Orange	Stump	Green
Orange	Orange	Orange	Orange	Orange	Stump	Orange
Green	Stump	Skeleton	Stump	Skeleton	Skeleton	Orange
Orange	Stump	Green	Green	Green	Orange	Orange
Orange	Skeleton	Green	Green	Skeleton	Green	Orange
Green	Orange	Skeleton	Orange	Green	Green	Green
Green	Orange	Green	Green	Green	Green	Green
Green	Green	Green	Orange	Green	Orange	Green
Green	Green	Green	Orange	Green	Green	Green
Green	Green	Orange	Orange	Orange	Skeleton	Skeleton
Orange	Orange	Stump	Skeleton	Stump	Green	Green
Skeleton	Green	Green	Green	Stump	Green	Orange
Green	Green	Green	Orange	Stump	Skeleton	Green
Orange	Orange	Orange	Green	Green	Orange	Green
Green	Green	Green	Green	Orange	Stump	Skeleton
Orange	Green	Orange	Green	Green	Skeleton	Green
Green	Green	Orange	Green	Orange	Skeleton	Green
Skeleton	Green	Green	Green	Green	Green	Green
Green	Orange	Orange	Green	Orange	Green	Green
Green	Green	Orange	Orange	Green	Green	Orange
Green	Orange	Skeleton	Skeleton	Orange	Skeleton	Stump
Skeleton	Orange	Green	Green	Stump	Orange	Stump
Orange	Green	Orange	Green	Stump	Stump	Orange
Stump	Green	Stump	Skeleton	Green	Stump	Green
Green	Green	Stump	Stump	Stump	Stump	Orange
Green	Green	Skeleton	Orange	Green	Skeleton	Green
Orange	Orange	Orange	Stump	Orange	Green	Skeleton
Green	Orange	Orange	Skeleton	Orange	Green	Stump
Green	Skeleton	Green	Orange	Orange	Green	Orange
Green	Green	Skeleton	Orange	Orange	Orange	Stump
Orange	Orange	Green	Green	Green	Green	Orange
Green	Orange	Green	Skeleton	Orange	Green	Orange
Orange	Green	Skeleton	Stump	Orange	Green	Orange
Green	Green	Orange	Skeleton	Orange	Orange	Orange
Orange	Green	Stump	Orange	Orange	Green	Stump
Green	Orange	Skeleton	Orange	Skeleton	Green	Orange
Green	Orange	Skeleton	Green	Skeleton	Orange	Green
Orange	Green	Green	Green	Green	Green	Green
Green	Green	Green	Green	Skeleton	Skeleton	Green

FIGURE 17.2 Piñon inventory, 2022. As Nina Elder embarks on sharing her life with a dying species, she will observe and catalog their journey. This color coded list signifies the state of each piñon on Nina's land at the end of 2022.

BIBLIOGRAPHY

Abram, David. *Becoming Animal: An Earthly Cosmology*. New York: Vintage, 2011.
Aloi, Giovanni. *Art and Animals*. London: I. B. Tauris, 2011.
Amstutz, Nina. "The Avian Sense for Beauty: A Posthumanist Perspective on the Bowerbird." *Art History* 44, no. 5 (2021): 1038–1064.
Armstrong, Rachel. "Self-Repairing Architecture." *NextNature* (blog), June 24, 2010. https://nextnature.net/story/2010/self%E2%80%93repairing-architecture.
Barad, Karen. *Meeting the Universe Halfway: Quantum Physics and the Entanglement of Matter and Meaning*. Durham, NC: Duke University Press, 2007.
Barnard, Timothy. *Nature's Colony: Empire, Nation and Environment in Singapore Botanic Gardens*. Singapore: NUS Press, 2017.
Battistoni, Alyssa. "Bring in the Work of Nature: From Natural Capital to Hybrid Labor." *Political Theory* 45, no. 1 (2017): 5–31.
Beatley, Timothy. *The Ecology of Place: Planning for Environment, Economy, and Community*. Washington, DC: Island Press, 1997.
Bedau, Mark A. "Living Technology Today and Tomorrow." *Technoetic Arts* 7, no. 2 (2009): 199–206.
Behar, Katherine, and Emmy Mikelson, eds. *And Another Thing: Nonanthropocentrism and Art*. Brooklyn: Punctum Books, 2016.
Bennett, Jane. *Vibrant Matter*. Durham, NC: Duke University Press, 2009.
Benyus, Janine. *Biomimicry: Innovation Inspired by Nature*. New York: Harper Collins, 1997.
Berry, Brian J. L. "Urbanization." In *The Earth as Transformed by Human Action: Global and Regional Changes in the Biosphere over the Past 300 Years*, edited by B. L. Turner, William C. Clark, Robert W. Kates, John F. Richards, J. T. Matthews and W. B. Meyer, 103–120. New York: Cambridge University Press, 1990.
Bjarke Ingels Group, ed. *Yes Is More: An Archicomic on Architectural Evolution*. Cologne: Taschen, 2009.
Bois, Yve-Alain, and Rosalind E. Krauss. *Formless: A User's Guide*. New York: Zone Books, 1997.
Borucke, Michael, David Moore, Gemma Cranston, Kyle Gracey, Katsunori Iha, Joy Larson, and Elias Lazarus et al. "Accounting for Demand and Supply of the Biosphere's Regenerative Capacity: The National Footprint Accounts' Underlying Methodology and Framework." *Ecological Indicators*, no. 24 (2013): 518–533.
Braidotti, Rosi. *The Posthuman*. Hoboken; Cambridge: Wiley; Polity Press, 2013.
Bratman, Eve Z., and William P. DeLince. "Dismantling White Supremacy in Environmental Studies and Sciences: An Argument for Anti-Racist and Decolonizing Pedagogies." *Journal of Environmental Studies and Sciences* 12, no. 2 (2022): 193–203.

Bravery, Benjamin D., James A. Nicholls, and Anne W. Goldizen. "Patterns of Painting in Satin Bowerbirds *Ptilonorhynchus violaceus* and Males' Responses to Changes in Their Paint." *Journal of Avian Biology* 37, no. 1 (2006): 77–83.

Brennan, Stella. "Border Patrol." In *Patricia Piccinini: In Another Life*. Wellington, NZ: Wellington City Gallery, 2006.

Buckley, Charles. *An Anecdotal History of Old Times in Singapore*. Vol. 1. Singapore: Fraser & Neave, 1902.

Burch, William R., Jo Ellen Force, and Gary E. Machlis. *The Structure and Dynamics of Human Ecosystems: Toward a Model for Understanding and Action*. New Haven: Yale University Press, 2018.

Cadenasso, Mary L., Anne M. Rademacher, and Steward T. Pickett. "Systems in Flames: Dynamic Coproduction of Social-Ecological Processes." *Bioscience* 72, no. 8 (2022): 731–744.

Carman-Brown, Kylie. "A Tale of Extremes." *The People & Environment Blog. National Museum of Australia*, April 7, 2015. https://pateblog.nma.gov.au/2015/04/07/a-tale-of-extremes/.

Cohen, Joel. "Population Growth and Earth's Carrying Capacity." *Science* no. 269 (1995): 341–346.

Cooper, Melinda. *Life as Surplus: Biotechnology and Capitalism in the Neoliberal Era*. Seattle: University of Washington Press, 2008.

Cox, Christoph. "Of Humans, Animals, and Monsters." Chap. 2 in *Becoming Animal: Contemporary Art in the Animal Kingdom*, edited by Nato Thompson. Cambridge, MA: MIT Press, 2005.

Cruz, Marcos. "Synthetic Neoplasms." *Architectural Design* 78, no. 6 (2008): 36–43.

Darwin, Charles. Charles Darwin to Asa Gray, April 3, 1869. In *Darwin Correspondence Project*, no. 2743. http://www.darwinproject.ac.uk/entry-2743.

Dawkins, Richard. *The Extended Phenotype: The Long Reach of the Gene*. Oxford: Oxford University Press, 1982.

Despret, Vinciane. *What Would Animals Say If We Asked the Right Questions?* Translated by Brett Buchanan. Minneapolis: University of Minnesota Press, 2016.

Diamond, Jared. "Animal Art: Variation in Bower Decorating Style among Male Bowerbirds, *Amblyornis inornatus*." *Proceedings of the National Academy of Sciences* 83, no. 9 (1986): 3042–3046.

———. "Experimental Study of Bower Decoration by the Bowerbird *Amblyornis inornatus*, Using Colored Poker Chips." *The American Naturalist* 131, no. 5 (1988): 631–653.

Dulhunty, Roma. *The Spell of Lake Eyre*. Kilmore, Vic: Lowden Publishing, 1975.

———. *When the Dead Heart Beats Lake Eyre Lives*. Kilmore, Vic: Lowden Publishing, 1979.

Edelman, Lee. *No Future: Queer Theory and the Death Drive*. Durham, NC: Duke University Press, 2004.

Eliasson, Olafur. "Why Art Has the Power to Change the World." *World Economic Forum*, January 18, 2016. https://www.weforum.org/agenda/2016/01/why-art-has-the-power-to-change-the-world/.

Ellis, Erle. *Anthropocene: A Very Short Introduction*. New York: Oxford University Press, 2018. Kindle Edition.

Ellis, Erle, and Navin Ramankutty. "Putting People on the Map: Anthropogenic Biomes of the World." *Frontier Ecological Environments* no. 6 (2008): 430–447.

Endler, John A., Lorna C. Endler, and Natalie R. Doerr. "Great Bowerbirds Create Theaters with Forced Perspective When Seen by Their Audience." *Current Biology* 20, no. 18 (2010): 1679–1684.

Endler, John A., Julie Gaburro, and Laura A. Kelley. "Visual Effects in Great Bowerbird Sexual Displays and Their Implications for Signal Design." *Proceedings: Biological Sciences* 281, no. 1783 (2014): 1–9.

Engberg, Juliana. "Atmosphere." In *Patricia Piccinini: Atmosphere, Autosphere, Biosphere*, edited by Juliana Engberg, Edward Colless, and Hiroo Yamagatam. Collingwood, Australia: Drome Pty Limited, 2000.

Er, Kenneth. "Transforming Singapore into a City in Nature." *Urban Solutions*, no. 19 (2021): 68–77. https://www.clc.gov.sg/docs/default-source/urban-solutions/urbsol19pdf/09_essay_transforming-singapore-into-a-city-in-nature.pdf.

Eyth, Max. *Lebendige Kräfte: Sieben Vorträge aus dem Gebiete der Technik*. 4th ed. Berlin: J. Springer (1905) 1924.

Fairs, Marcus. "Mycelium Chair by Eric Klarenbeek is 3D-Printed with Living Fungus." *deezen*, October 20, 2013. https://www.dezeen.com/2013/10/20/mycelium-chair-by-eric-klarenbeek-is-3d-printed-with-living-fungus/.

FEMA. *Building Community Resilience with Nature-Based Solutions: A Guide for Local Communities*. Washington, DC: FEMA, 2021. https://www.fema.gov/sites/default/files/documents/fema_riskmap-nature-based-solutions-guide_2021.pdf.

Fisch, Michael. "The Nature of Biomimicry: Toward a Novel Technological Culture." *Science, Technology & Human Values* 42, no. 5 (2017): 795–821.

Fischer, Anke, and Antonia Eastwood. "Coproduction of Ecosystem Services as Human-Nature Interactions—An Analytical Framework." *Land Use Policy* 52 (2016): 41–50.

Frith, Clifford, and Dawn Frith. *The Bowerbirds Ptilonorhynchidae*. Oxford: Oxford University Press, 2004.

Gat, Orit. "Global Audiences, Zero Visitors: How to Measure the Success of Museums' Online Publishing." *Rhizome*, March 13, 2015. http://rhizome.org/editorial/2015/mar/12/global-audiences-zero-visitors/.

Gelfand, Janelle. "Inside a Symphony Audition." *Cincinnati Enquirer*, November 24, 2015. https://www.cincinnati.com/story/entertainment/music/2015/11/24/inside-symphony-audition/75478764/.

Giffney, Noreen, and Myra Hird, eds. *Queering the Non/Human*. Farnham, UK: Ashgate, 2008.

Gray, Denis. "Cambodia Sells Sand; Environment Ravaged." *Asian Reporter* 21, no. 17 (2011): 4.

Gregory, J. W. *The Dead Heart of Australia: A Journey Around Lake Eyre in the Summer of 1901-1902, With Some Accounts of the Lake Eyre Basin and the Flowing Wells of Central Australia*. London: Murray, 1906.

Hammound, Ryan, Stefania Tognin, Lucie Burgess, Nicol Bergou, Michael Smythe, Johanna Gibbons, Neil Davidson, Alia Afifi, Ioannis Bakolis, and Andrea Mechelli. "Smartphone-Based Ecological Momentary Assessment Reveals Mental Health Benefits of Birdlife." *Scientific Reports* 12, no. 1 (2022). https://doi.org/10.1038/s41598-022-20207-6.

Harari, Yuval Noah. *Sapiens: A Brief History of Humankind*. New York: Harper Collins, 2015.

Haraway, Donna. "A Manifesto for Cyborgs: Science, Technology, and Socialist Feminism in the 1980s." In *Simians, Cyborgs and Women*, 149–181. New York: Routledge, 1991.

_____. "Situated Knowledges: The Science Question in Feminism and the Privilege of Partial Perspective." *Feminist Studies* 14, no. 3 (1988): 575–599.

_____. "Speculative Fabulations for Technoculture's Generations." In *(Tender)Creatures*. Vitoria-Gasteiz, Spain: Atrium Gallery, 2007.

_____. *Staying with the Trouble: Making Kin in the Chthulucene*. Durham NC: Duke University Press, 2016.

_____. *When Species Meet*. Minneapolis: University of Minnesota Press, 2008.

Harjo, Joy. *She Had Some Horses: Poems*. New York: W. W. Norton, 2008.

Harris, Daniel. *Cute, Quaint, Hungry, and Romantic: The Aesthetics of Consumerism*. New York: Basic Books, 2000.

Harrison, Ariane Lourie, ed. *Architectural Theories of the Environment: Posthuman Territory*. New York: Routledge, 2012.

Hazen, Robert. *The Story of the Earth: The First 4.5 Billion Years, from Stardust to Living Planet*. New York: Penguin Books, 2012.

Hershberger, Scott. "Humans Are All More Closely Related Than We Commonly Think." *Scientific American*, October 5, 2020. https://www.scientificamerican.com/article/humans-are-all-more-closely-related-than-we-commonly-think/.

Hervé-Gruyer, Charles, and Perrine Hervé-Gruyer. *Permaculture: guérir la terre, nourrir les hommes*. Paris: Actes du Sud, 2014.

Hovorka, Alice. "Transspecies Urban Theory: Chickens in an African City." *Cultural Geographies* 15, no. 1 (2008): 95–117.

Huiwen, Ng. "National Day Rally 2019: 8 Things to Know about PM Lee Hsien Loong's Speech." *Straits Times*, October 26, 2019. https://www.straitstimes.com/politics/national-day-rally-2019-8-things-to-know-about-pm-lee-hsien-loongs-speech.

Hurley, Amandy Kolson. "Floating Cities Aren't the Answer to Climate Change." *Bloomberg CityLab*, April 19, 2019. https://www.bloomberg.com/news/articles/2019-04-10/floating-cities-won-t-save-us-from-climate-change.

Intergovernmental Science-Policy Platform on Biodiversity and Ecosystem Service (IPBES). *Summary for Policymakers of the Global Assessment Report on Biodiversity and Ecosystem Services of the Intergovernmental Science-Policy Platform on Biodiversity and Ecosystem Services*, edited by S. Díaz, J. Settele, E. S. Brondízio, H.T. Ngo, M. Guèze, J. Agard, A. Arneth et al. Bonn: IPBES Secretariat, 2019.

IPCC. *Climate Change 2022: Impacts, Adaptation, and Vulnerability*, edited by Hans-Otto Pörtner, Debra C. Roberts, Melinda M. B. Tignor, Elvira Poloczanska, Katja Mintenbeck, Andrés Alegría, Marlies Craig et al. New York: Cambridge University Press, 2002. https://doi.org/10.1017/9781009325844.

Jamieson, William. "There's Sand in My Infiniti Pool: Land Reclamation and the Rewriting of Singapore." *GeoHumanities* 3, no. 2 (2017): 396–413. https://doi.org/10.1080/2373566X.2017.1279021.

Kant, Immanuel. *Fundamental Principles of the Metaphysic of Morals*. Translated by T. K. Abbott. New York: Prometheus, 1988.

Kelley, Laura A., and John A. Endler. "How Do Great Bowerbirds Construct Perspective Illusions?" *Royal Society Open Science* 4, no. 1 (2017): 1–10.

Knoll, Andrew. *A Brief History of Earth: Four Billion Years in Eight Chapters*. New York: Harper Collins, 2021.

Krough, Anders. *State of the Tropical Rainforest: The Complete Overview of the Tropical Rainforest, Past and Present*. Oslo: Rainforest Foundation Norway, 2020. https://d5i6is0eze552.cloudfront.net/documents/Publikasjoner/Andre-rapporter/RF_StateOfTheRainforest_2020.pdf?mtime=20210505115205.

Kurmanaev, Anatoly. "A Battle of Singing Stars, with Wings and Feathers." *New York Times*, January 14, 2021. https://www.nytimes.com/2021/01/14/world/americas/suriname-birds.html.

Lange-Berndt, Petra. *Materiality (Documents of Contemporary Art)*. Cambridge, MA: MIT Press, 2015.

Larson, Brendon. *Metaphors for Environmental Sustainability: Redefining Our Relationship with Nature*. New Haven: Yale University Press, 2011.

Latour, Bruno. *Politics of Nature: How to Bring Sciences into Democracy*. Cambridge: Harvard University Press, 2004.

Lee, Kuan Yew. *From Third World to First: The Singapore Story, 1965-2000*. Vol. 2, *Memoirs of Lew Kuan Yew*. Singapore: Singapore Press Holdings, 2008.

Lewis, Sarah Elizabeth. *The Rise: Creativity, the Gift of Failure, and the Search for Mastery*. New York: Simon & Schuster, 2014.

Light, Jennifer S. *The Nature of Cities: Ecological Visions and the American Urban Professions, 1920–1960*. Baltimore: Johns Hopkins University Press, 2009.

Lim, Tin Seng. "Land from Sand: Singapore's Reclamation Story." *Biblioasia* 13, no. 1 (04 April 2017). https://biblioasia.nlb.gov.sg/vol-13/issue-1/apr-jun-2017/land-from-sand#fn:1.

Longino, Helen E. *Science and Social Knowledge: Values and Objectivity in Scientific Inquiry*. Princeton: Princeton University Press, 1990.

Machlis, Gary E., Miguel O. Román, and Steward T. Pickett. "A Framework for Research on Recurrent Acute Disasters." *Science Advances* 8, no. 10 (2022). https://doi.org/10.1126/sciadv.abk2458.

Margulis, Lynn. *Symbiotic Planet: A New Look at Evolution*. New York: Basic Books, 1998.

Massumi, Brian. *Parables for the Virtual: Movement, Affect, Sensation*. Durham NC: Duke University Press, 2002.

McDonald, Helen. *Patricia Piccinini: Nearly Beloved*. Dawes Point, N. S. W: Piper Press, 2012.

Media press release for *Patricia Piccinini: Relativity* (Perth: Art Gallery of Western Australia, 2010). http://www.artgallery.wa.gov.au/about_us/documents/Patricia-Piccinini-media-release-2010.pdf.

Meiners, Scott J., Steward T. Pickett, and Mary L. Cadenasso. *An Integrative Approach to Successional Dynamics: Tempo and Mode of Vegetation Change*. New York: Cambridge University Press, 2015.

Michael, Linda. *Patricia Piccinini: We Are Family*. Strawberry Hills, NSW: Australia Council, 2003. Exhibition Catalog for the Australia Pavilion at the Venice Biennale.

Mildenberger, Matto. "The Tragedy of the *Tragedy of the Commons*." *Scientific American* (blog), April 23, 2019. https://blogs.scientificamerican.com/voices/the-tragedy-of-the-tragedy-of-the-commons/.

Mitsch, W. J. "What Is Ecological Engineering?." *Ecological Engineering* 45 (2012): 5–12.

Mondrian, Piet. *Natural Reality and Abstract Reality: An Essay in Trialogue Form in 1919–1920*. New York: George Brazilier, 1995.

Moran, Joe. "Earthrise: The Story behind Our Planet's Most Famous Photo." *The Guardian*, December 22, 2018. https://www.theguardian.com/artanddesign/2018/dec/22/behold-blue-plant-photograph-earthrise.

Murdoch, Lindsay. "Sand Wars: Singapore's Growth Comes at the Environmental Expense of its Neighbours." *Sydney Morning Herald*, February 26, 2016.

Naess, Arne. "The Shallow and the Deep, Long Range Ecology Movement: A Summary." *Inquiry: An Interdisciplinary Journal of Philosophy* 16, no. 1–4 (1973): 95–100.

Nagel, Thomas. "What Is It Like to Be a Bat?" *The Philosophical Review* 83, no. 4 (1974): 435–450.

Nishime, Leilani, and Kim D. Hester Williams, eds. *Racial Ecologies*. Seattle: University of Washington Press, 2018.

O'Neil, Daniel, Andrew Fanning, William Lamb, and Julia Steinberger, "A Good Life for all Within Planetary Boundaries." *Nature Sustainability* 1 (2018): 88–95. https://doi.org/10.1038/s41893-018-0021-4.

Otter, Ken A., Alexandra McKenna, Stefanie E. LaZerte, and Scott M. Ramsay. "Continent-wide Shifts in Song Dialects of White-Throated Sparrows." *Current Biology* 30, no. 16 (2020): 3231–3235.

Palmer, Matthew, Kathleen McInnes, and Mohar Chattopadhyay. "Key Factors for Sea Level Rise in the Singapore Region." In *Singapore 2nd National Climate Change Study*, Phase 1. Singapore, 2015. https://doi.org/10.13140/RG.2.1.1875.9449.

Park, Robert E., and Ernest W. Burgess. *The City*. Chicago: University of Chicago Press, 2019.

Parr, Adrian. "Commonism." In *Birth of a New Earth: The Radical Politics of Environmentalism*, 91–119. New York: Columbia University Press, 2018.

Pentecost, Claire. "Outfitting the Laboratory of the Symbolic: Toward a Critical Inventory of Bioart." In *Tactical Biopolitics: Art, Activism and Technoscience*, edited by Beatrice da Costa and Kavita Philip, 107–123. Cambridge, MA: MIT Press, 2008.

Pickering, Andrew. "Asian Eels and Global Warming: A Posthumanist Perspective on Society and the Environment." *Ethics & Environment* 10 (2005): 29–43.

Pickett, Steward T., and Mary L. Cadenasso. "The Ecosystem as a Multidimensional Concept: Meaning, Model, and Metaphor." *Ecosystems* 5, no. 1 (2022): 1–10.

Piccinini, Patricia. "Patricia Piccinini in Conversation with Alasdair Foster." *Photofile* 68 (2003): 22.

_____. *Public Presentation at the Tokyo National University of Fine Arts and Music*. Tokyo, Japan: Faculty of Fine Arts, December 8, 2003.

Pickett, Steward T., Mary L. Cadenasso, M. Anne, and Rademacher. "Toward Pluralizing Ecology: Finding Common Ground Across Sociocultural and Scientific Perspectives." *Ecosphere* 13, no. 9 (2022): e4231. https://doi.org/10.1002/ecs2.4231.

Pickett, Steward T., Mary L. Cadenasso, and J. M. Grove. "Resilient Cities: Meaning, Models, and Metaphor for Integrating the Ecological, Socio-Economic, and Planning Realms." *Landscape and Urban Planning* 69, no. 4 (2004): 369–384.

Pickett, Steward T., V. Thomas Parker, and Peggy L. Fielder. "The New Paradigm in Ecology: Implications for Conservation Biology Above the Species Level." In *Conservation Biology: The Theory and Practice of Nature Conservation, Preservation, and Management*, edited by Peggy L. Fiedler and Subodh K. Jain, 65–88. New York: Chapman and Hall, 1992. https://doi.org/10.1007/978-1-4684-6426-9_4.

Pievani, Telmo. "The Sixth Mass Extinction: Anthropocene and the Human Impact on Biodiversity." *Rendiconti Lincei*, no. 25 (2014): 85–93. https://doi.org/10.1007/s12210-13-0258-9.

Poe, Edgar Allan. *The Fall of the House of Usher and Other Tales*. New York: Penguin Random House, 2006.

Polidoro, Beth, Kent Carpenter, Lorna Collins, Norman Duke, Aaron Ellison, and Joanna Ellison et al. "The Loss of One Species: Mangrove Extinction Risk and Geographic Areas of Global Concern." *PLoS ONE* 5, no. 4 (2010): e10095. https://doi.org/10.1371/journal.pone.0010095.

Powell, Miles Alexander. "Singapore's Lost Coast: Land Reclamation, National Development and the Erasure of Human and Ecological Communities, 1822–Present." *Environment and History* 27, no. 4 (2021): 635–663. https://doi.org/10.3197/096734019X15631846928710.

Prum, Richard. *The Evolution of Beauty: How Darwin's Forgotten Theory of Mate Choice Shapes the Animal World—and Us*. New York: Random House, 2017.

Public Works Department. *Annual Report*. Singapore: Public Works Department, 1975. National Library Board of Singapore.

Quirk, Joe, and Patri Friedman. *Seasteading: How Floating Nations Will Restore the Environment, Enrich the Poor, Cure the Sick, and Liberate Humanity from Politicians*. New York: Free Press, 2017.

Rademacher, Anne, Mary L. Cadenasso, and Steward T. Pickett. "From Feedbacks to Coproduction: Toward an Integrated Conceptual Framework for Urban Ecosystems." *Urban Ecosystems* 22, no. 1 (2019): 65–76.

Rampley, Matthew. *The Seductions of Darwin: Art, Evolution, Neuroscience*. University Park: Pennsylvania State University Press, 2017.

Retallack, Joan. "The Ventriloquist's Dilemma." In *Bosch'd*. Brooklyn: Litmus Press, 2020.

———. *The Poethical Wager*. Berkeley: University of California Press, 2003.

Roeske, Tina, David Rothenberg, and David Gammon. "Mockingbird Morphing Music: Structured Transitions in a Complex Bird Song." *Frontiers in Psychology* 12 (2021). https://doi.org/10.3389/fpsyg.2021.630115.

Rose, Deborah Bird. *Reports from a Wild Country: Ethics for Decolonisation*. Sydney: University of New South Wales Press, 2004.

Rothenberg, David. *Survival of the Beautiful: Art, Science, and Evolution*. London: Bloomsbury, 2011.

———. *Why Birds Sing: A Journey into the Mystery of Bird Song*. New York: Basic Books, 2005.

Schell, Christopher J., Karen Dyson, Tracy L. Fuentes, Simone Des Roches, Nyeema C. Harris, Danica Sterud Miller, Cleo A. Woelfe-Erskine, and Max R. Lambert. "The Ecological and Evolutionary Consequences of Systemic Racism in Urban Environments." *Science* 369, no. 6510 (2020). https://doi.org/10.1126/science.aay4497.

Senanayake, Ranil, and John Jack. *Analogue Forestry: An Introduction*. Clayton: Department of Geography and Environmental Science at Monash University, 1998.

Seymour, Nicole. *Strange Natures: Futurity, Empathy, and the Queer Ecological Imagination*. Urbana: University of Illinois Press, 2013.

Singer, Peter. *Animal Liberation: A New Ethics for Our Treatment of Animals*. New York: Random House, 1975.

Smil, Vaclav. *Enriching the Earth: Fritz Haber, Carl Bosch, and the Transformation of World Foods*. Cambridge: MIT Press, 2001.

Spencer, Douglas. "Island Life: The Eco-Imaginary of Capitalism." *Log* no. 47 (2019): 167–174.

Stalter, Richard, Jingjing Tong, and James Lendemer. "The Flora on the High Line, New York City, New York: A 17-Year Comparison." *The Journal of the Torrey Botanical Society* 148, no. 3 (2021): 243–251. https://doi.org/10.3159/TORREY-D-21-00007.1.

Stein, Gertrude. *What are Masterpieces?* New York: Pitman Pub. Corp, 1970. Accessed on Internet Archive.

Steyerl, Hito. "In Defense of the Poor Image." *e-Flux* 10 (November 2009), http://www.e-flux.com/journal/10/61362/in-defense-of-the-poor-image/.

Straits Times. "New Airport is On Site of Former Swamp." June 12, 1937. National Library Board Singapore.

Strand, Sophie. "Rewilding Mythology." *Atmos*, September 15, 2022. https://atmos.earth/rewilding-mythology-mythmaking-climate-collapse/.

Strassburg, Bernardo B. N., Alvaro Iribarrem, Hawthorne L. Beyer, Carlos L. Cordeiro, Renato Crouzeilles, Catarina C. Jakovac, and André B. Junqueira et al. "Global Priority Areas for Ecosystem Restoration." *Nature* 586 (2020): 724–729. https://doi.org/10.1038/s41586-020-2784-9.

Taylor, Hollis. *Is Birdsong Music? Outback Encounters with an Australian Songbird*. Bloomington: Indiana University Press, 2017.

Temür, Başak Doğa. "Just Become Something Is Bad, Doesn't Mean It Isn't Good." In *Hold Me Close to Your Heart*. Istanbul: ARTER, 2011. http://www.patriciapiccinini.net/printessay.php?id=37.

Thacker, Eugene. *Global Genome: Biotechnology, Politics, and Culture*. Cambridge, MA: MIT Press, 2005.

The Economist. "Such Quantities of Sand; Banyan." February 28, 2015.

Truth and Reconciliation Commission of Canada. *Honouring the Truth, Reconciling for the Future: Summary of the Final Report of the Truth and Reconciliation Commission of Canada*. Ottawa: Truth and Reconciliation Commission of Canada, 2015. https://publications.gc.ca/site/eng/9.800288/publication.html.

Tuin, Iris van der, and Rick Dolphijn, eds. "Interview with Karen Barad." In *New Materialism Interviews & Cartographies*. Ann Arbor: Open Humanities Press, 2012.

Ullrich, Jessica. "Jedes Tier ist eine Künstlerin." In *Das Handeln der Tiere – Tierliche Agency im Folkus der Human-Animal Studies*, edited by Sven Wirth, Anett Laue, Markus Kurth, Katharina Dornenzweig, Leonie Bossert and Karsten Balgar, 245–265. Bielefeld: De Gruyter, 2015.

United Nations. "Introduction." In *The World at Six Billion*, 1–3. United Nations: 1999. https://www.un.org/development/desa/pd/sites/www.un.org.development.desa.pd/files/files/documents/2020/Jan/un_1999_6billion.pdf.

Uy, J. Albert C., and Gerald Borgia. "Sexual Selection Drives Rapid Divergence in Bowerbird Display Traits." *Evolution* 54, no. 1 (2000): 273–278.

Vitousek, Peter M. "Human Domination of Earth's Ecosystems." *Science* 277 (1997): 494–499.

Von Uexküll, Jakob. "A Stroll Through the World of Animals and Men: A Picture Book of Invisible Worlds." *Semiotica* 89, no. 4 (1992): 319–391.

Vuong, Ocean. *On Earth We're Briefly Gorgeous: A Novel*. London: Penguin Press, 2021.

Watkins, Mary, and G. A. Bradshaw. "Trans-Species Psychology: Theory and Praxis." *Spring: A Journal of Archetype and Culture* 76 (2007): 69–94.

Weathers, Kathleen C., David L. Strayer, and Gene E. Likens, eds. *Fundamentals of Ecosystem Science*, 2nd ed. London: Academic Press, 2021.

Williams, Linda. "Spectacle or Critique? Reconsidering the Meaning of Reproduction in the Work of Patricia Piccinini." *Southern Review: Communication, Politics, Culture* 37, no. 1 (2004): 76–94.

Wolch, Jennifer R., Jason Byrne, and Joshua P. Newell. "Urban Green Space, Public Health, and Environmental Justice." *Landscape and Urban Planning* 125 (2014): 234–244.

Wolch, Jennifer R., Kathleen West, and Thomas E. Gaines. "Transspecies Urban Theory." *Environment and Planning D: Society and Space* 13, no. 6 (1995): 735–760.

Wolfe, Cary. *Animal Rites: American Culture, the Discourse of the Species and Posthumanist Theory*. Chicago: University of Chicago Press, 2003.

———. *Before the Law: Humans and Other Animals in a Biopolitical Frame*. Chicago: University of Chicago Press, 2013.

Wu, Jianguo, and Orie L. Loucks. "From Balance of Nature to Hierarchical Patch Dynamics: A Paradigm Shift in Ecology." *The Quarterly Review of Biology* 70, no. 4 (1995): 439–466.

Yeoh, Brenda. *Contesting Space in Colonial Singapore: Power Relations and the Urban Built Environment*. Singapore: NUS Press, 2003.

Zylinkska, Joanna. *Bioethics in the Age of New Media*. Cambridge, MA: MIT Press, 2009.

Websites

Boeri: Stefano Boeri Architects. "Vertical Forest." Accessed March 2, 2023. https://www.stefanoboeriarchitetti.net/en/project/vertical-forest/.

Channel News Asia. "Lost Waterfronts." Uploaded February 13, 2023, Singapore. Video, 47:34. https://www.channelnewsasia.com/watch/lost-waterfronts.

Gardens by the Bay. "Super Tree Grove." Accessed May 14, 2023. https://www.gardensbythebay.com.sg/en/things-to-do/attractions/supertree-grove.html.

Global Footprint Network. "Earth Overshoot Day." Accessed January 6, 2023. https://www.overshootday.org/.

———. "Data and Methodology." Accessed January 7, 2023. https://www.footprintnetwork.org/resources/data/.

———. "How Many Earths? How Many Countries?" Accessed January 6, 2023. https://www.overshootday.org/how-many-earths-or-countries-do-we-need/.

Marx, Karl. "Capital: A Critique of Political Economy, Volume 1." Marxist Internet Archive. Accessed October 18, 2022. https://www.marxists.org/archive/marx/works/1867-c1/ch07.html.

National Parks Singapore. "Singapore, Our City in Nature." Last updated January 2023. https://www.nparks.gov.sg/about-us/city-in-nature.

NUS Deltares. "Polder Development at Pulau Tekong." Accessed May 14, 2023. https://nusdeltares.info/projects/project-2/.

Patricia Piccinini. "Artist's Statement on *SO2*." Accessed November 10, 2015. http://www.patriciapiccinini.net.

Poetry Foundation. "Hurt Hawks by Robinson Jeffers." Accessed May 16, 2023. https://www.poetryfoundation.org/poems/51675/hurt-hawks.

———. "Robinson Jeffers." Accessed January 3, 2023. https://www.poetryfoundation.org/poets/robinson-jeffers.

PUB, Singapore's National Water Agency. "Sea Level Rise." Last modified October 13, 2022. https://www.pub.gov.sg/Pages/sealevelrise.aspx.

Roots. "Beach at the Foot of Mount Palmer." Last modified April 2, 2021. https://www.roots.gov.sg/Collection-Landing/listing/1142729.

Shorter, Wayne. "The Language of the Unknown." Uploaded on December 1, 2016. Video, 56:56. https://www.youtube.com/watch?v=sy17GpcZ79w.

Singapore Government Agency. "Green Plan 2030." Last updated May 4, 2023. https://www.greenplan.gov.sg/.

Singapore National Parks. "Intertidal." Last updated January 2023. https://www.nparks.gov.sg/biodiversity/our-ecosystems/coastal-and-marine/intertidal.

———. "Mangroves." Last updated January 2023. https://www.nparks.gov.sg/biodiversity/our-ecosystems/coastal-and-marine/mangroves.

———. "Singapore, Our City in Nature." Last updated January 2023. https://www.nparks.gov.sg/about-us/city-in-nature.

Turtle Island Restoration Network. "How to Grow a Redwood Tree from Seed." Accessed January 3, 2023. https://seaturtles.org/how-to-grow-a-redwood-tree-from-seed/.

United Nations. "Day of 8 Billion." Accessed January 4, 2023. https://www.un.org/en/dayof8billion#:~:text=On%2015%20November%202022%2C%20the,nutrition%2C%20personal%20hygiene%20and%20medicine.

Wikipedia. "Land Reclamation in Singapore." Last modified February 10, 2023. https://en.wikipedia.org/wiki/Land_reclamation_in_Singapore.

———. "Olafur Eliasson." Accessed December 18, 2022. https://en.wikipedia.org/wiki/Olafur_Eliasson.

Wild Singapore. "Mud Lobster." Last updated March 2020. http://www.wildsingapore.com/wildfacts/crustacea/othercrust/lobster/thalassina.htm.

World Bank. "Urban Development." Accessed January 24, 2023. https://www.worldbank.org/en/topic/urbandevelopment/overview.

INDEX

Note: Page references in *italics* denote figures, and with "n" endnotes.

2001 Roadless Rule 132

Abbey, Edward 127
abyss: ancestral cells in 84; and chemical chain reactions 83; life appearing in 83
"adorable mutant infant" 59
advocacy 116–117
aesthetic force 138–139
agential realism 28
agriculture: biotechnology 17; industrialized animal 129; regenerative 131
algaes 86–87; evolutions of 87–88; extinctions of 87–88
ancestral cells 83–84
Anders, William 138
"animalcules" 81
Animal Influencers 112–113
animal music 4
Animal Rites (Wolfe) 127, 128
animals 86–87; aesthetics 103; ambassadors 113; communication 108–109; evolutions of 87–88; extinctions of 87–88
The Animal That Therefore I Am (Derrida) 129
Ant Farm 51
anthromes 17
Anthropocene 107–108
anthropocentric activities 2, 16
anthropogenic biomes 17
anthropomorphism 57, 146
Ants of the Prairie 4
archaea 83–85
architect as advocate 116
"architects-in-residence" program 47
Architectural Ecologies Lab 52

architecture 22, 116, 118–120; coastal 41; design tactics for 33; human 28, 122; landscape 129, 131, 134; municipal 35
Armstrong, Rachel 21–22
art 141–142, *143*
artificial intelligence (AI) 109–110, 111–112
artists work 140–141
Austin Bat Refuge 109

bacteria 83–85
Barad, Karen 28, 66
Bardo (Tibetan Buddhism) 138
Barnard, Timothy 43
Bat Cloud 120
beaches 129–130
beach line 130
Becoming Animal (Abram) 56
Bedau, Mark 21–22
"Bee vase" 110
Before the Law (Wolfe) 128
Bengtson, Carla 4, 144–149
BIG 52
bioart 56–67; defined 56; objects 57, 64–65
biocapacity 16–17
biodiversity: ancestral cell 83–84; bacteria/archaea 83–85; brief history of life 82; building blocks of life 81–82; and ecosystem 44; and emotional wellbeing 2; evolutions/extinctions of animals/plants/algaes/fungi 87–88; intertidal 48; LECA 85–86; life appearing in abyss 83; LUCA 83–84; origin/evolution of 81–89; protists 84–85; protists/pluricellular organisms 86–87; scenarios for origin of life 83
Biodiversity Research Station 148

biofouling 52
biogeography 9
biomaterials 33–35
biomes: anthropogenic 17; defined 17
biomimetic design 29–30
biomimicry 3, 18–19; biomimetic design 29–30; and transpecies design 24–31
Biosphere 2 119
biotechnologies 57, 59; agriculture 17; biotechnological reproduction 62
biotic diversity 12
Bird, Deborah 68n2
bird-glass collision 116
birdsong 93–103
Bodyguards/Nature's Little Helpers 63
Boeri, Stefano 21
Bombus impatiens 19
Bower project 119, *119*
Braidotti, Rosi 65–66, 68n6, 70n38, 128, 129, 132
The Breathing Room 62
Brecht, Bertolt 149
Brennan, Stella 62, 69n35
Buffalo Audubon Society 120
building blocks of life 81–82
Buoyant Ecologies Float Lab 3, 52, *53*, 53–55
Butler, Octavia 149

Canadian Truth and Reconciliation Commission states 21
Cape Cod beaches 129
Carlyle, Warren 112
Catts, Oron 68n14
cells 81; eukaryotic 84–85; evolution *89*
charisma 118
Chek Jawa Wetlands 46–47
Chthulucene project *107*, 107–108
"City in Nature" framework (Singapore) 40–42, 45
climate change 138–139, 142, 147
co-design 32n22, 113
co-designers 3, 26, 30, 32n22
colonialism 12; Singapore 42
Coltrane, John 96
commensal 127; communities 4, 127; networks 129; spaces 134; species 129
commensalist projects 133
condor hatchling 128
Cousteau, Jacques 112
Cox, Christoph 56
(Tender) Creatures 60
Cruz, Marcos 68n14
Cummins Foundation Program 122
Cute, Quaint, Hungry, and Romantic: The Aesthetics of Consumerism (Harris) 63
"cutely grotesque" aesthetic 63
"cutely grotesque" creatures 57
Cuvier, Georges 87

Dark Side of the Moon (Floyd) 96
Darwin, Charles 81, 93–95, 97
Dead Heart (Gregory) 71
decentering 9; in ecology and transpecies design 13–14; people 13–14; in science of ecology 12
"deep listening" 110
deforestation 91; rates 1
de Kooning, Willem 30–31
Derrida, Jacques 129–130, 131
Descartes, René 111
The Descent of Man (Darwin) 94
Dickey, James 138
Dicks, Henry 2–3, 18–19
Dirt Witches 91
diversity: biotic 12; geologic 12; human 13
Dolphin Embassy 51
"Dreams of Deep Ecology" (Luke) 128
Dulhunty, John 71, 73
Dulhunty, Roma 71, 73

Earth Overshoot Day 16
Earthrise (Anders) 138
Earth Species Project 112
ecological footprint 16
ecological science: transformation 10; and transpecies design 9–14
ecology/ies 9–10, 13, 155; contemporary science of 9–10; as metaphor 11–12; Western science of 9
Elder, Nina 4
Elevator B 120
Eliasson, Olafur 139, 142
Elliot, T.S. 92
Ellis, Earl 17
El Niño Southern Oscillation 113
Emel, Jody 2
Engberg, Juliana 60
Engineered Living Materials (ELMs) 33–35; collaboration and outreach 36, *37–38*; objectives of 35; research activities 36
entangled intelligences 105–113; animal influencers 112–113; *Chthulucene* project *107*, 107–108; glaciers 110; nocturnal fugue 108–109; role of AI/technology 111–112; squeeker (mouse coach) 108, *109*; squid map 105–106, *106*; transpecies co-creation 110–111
entanglements 120, 123
ethics 66–67
Euglossa sp. 144
eukaryotic cells 84–85
evolutions: of algaes/fungi 87–88; of animals 87–88; of plants 87–88
Exhibit Columbus Project 116, 122, 124, 125
extinctions: of algaes/fungi 87–88; of animals 87–88; of plants 87–88
Eyth, Max 31

fence lizards 146–147
Fisch, Michael 28–29
Floyd, Pink 96
FluentPet 112
Foldscope 112
Fordham Press 129
Friedman, Milton 51
Friedman, Patri 51
Frontiers in Psychology 97
"full responsibility" 127–128, 131, 132
fungi 86–87; evolutions of 87–88; extinctions of 87–88

Game Boys Advanced 58
Gammon, David 96–97
"Garden City" vision of Singapore 44–45
geological time 73–74
geologic diversity 12
Germany 115, 123
gestural language 146
Ghoche, Ralph 4, 115–126
Ginsberg, Alexandra Daisy 112
glaciers 4, 110, 154
Glass, Philip 149
Glass Orchestras 137
Global Footprint Network 16–17
Google Arts and Culture Lab 112
Grandin, Temple 123
Gray, Asa 93
Great Artesian Basin 71, 73–74
"Great Oxygenation Event" (GOE) 84, 86–87
Gregory, John Walter 71

habitats: conducive structures 122; ecological coastal 3; for Griffis Sculpture Park 120; natural/unnatural 120; rich ecological 125
Haeckel, Ernst 81
Hanstholm summer LandShape Festival 133
Haraway, Donna 2, 20, 62, 63, 64–65, 68n2, 128–129
Hardin, Garrett 11
Häring, Hugo 123
Harjo, Joy 141
Harris, Catherine Page 4
Harris, Daniel 63
Hawaiian Bobtail Squid 106
hedonistic sustainability 52
heterogenous multiplicity of living 127, 131, 132
Hold Me Close to Your Heart exhibition 60
Homo sapiens 1–2, 16–18; vs. *Bombus impatiens* 19
Hooke, Robert 81
Hultsch, Henrike 100
human(s): as audiences 118; child 128; in commensal community 127; exceptionalism and bounded individualism 2, 4, 20; language 146; long-distance travel 130; nature/culture 145; population growth 16–17; relationship between non-human and 128; sentience 128; spatial landscape 130; well-being of 129
human activities: and earth's biodiversity/ecological processes 17; and extinction of species 1; and Holocene extinction 88
humanism: traditional 64; unbecoming 63–64
"Hurt Hawks" (Jeffers) 127
Hutto, Joe 132
Huxley, Julian 118
Hwang, Joyce 4, 115–126

Ikeda, Margaret 3
I'm Lost in Paris (Roche) 122
iNaturalist 112
industrialized animal agriculture 129
Industrial Revolution 107
information 10, 13, 90, 139–140; -carrying molecules 83–84; transfer of 71
inhumanism 128
Inomata, Aki 111
intensive spaces of becoming 127, 128, 132
Intergovernmental Panel on Climate Change (IPCC) 41
interspecies design 2, 17–20
intertidal biodiversity 48
invasive species 1, 5n1
ips confusus 154
"Is Looking at Art a Path to Mental Well-Being?" 141

Jeffers, Robinson 127–128, 132
Jensen, Derrick 132
Joachim, Mitchell 3
Jones, Evan 3
juniper-piñon ecosystem 151
Jüssi, Fred 101

Kaprielian, Gabriel 3
Kati Thanda (Lake Eyre), Australia 3, 71–75
Kenny G 96
Kewalo Marine Biology Lab in Hawaii 105
Kimmerer, Robin Wall 141
Klarenbeek, Eric 21
Kreysler & Associates 52

land reclamation 42–44, 51
landscapes 129; architecture 129, 131, 134; human spatial 130; non-human spatial 130; no-till 131; process of traversing 136; seed-established 131; smooth surfaces in 131; urban 130
Lange-Berndt, Petra 65
La Niña weather patterns 72
Laurence, Janet 3, 90–92
Leather Landscape 58, 58–59, 69n35
LECA ("Last Eukaryotic Cell Ancestor") 85–86

Lee Hsien Loong 41
LeRoux, Darren 124
Lewis, Sarah Elizabeth 138
Lewis Tsurumaki Lewis 47–48
Li, Jiabao 4
life: appearing in abyss 83; brief history of 82; building blocks of 81–82; microbial 74–75; scenarios for the origin of 83
lifeboat Earth metaphor 11
Life magazine 138
Life Support project 124
liminality 4, 41, 146
lizard language 146–147
Lubetkin, Berthold 118
LUCA (Last Universal Common Ancestor) 82–85
Luke, Tim 127, 128

MacKinnon, J. B. 128, 129
mantis shrimp 140
Marcus, Adam 3
Martins, Emilia 146–147
Marx, Karl 26
Massumi, Brian 76
McDonald, Helen 60, 64, 70n44
Mellor Primary School 115
melodiousness, defined 95
Merlin Tuttle's Bat Conservation 112
metabolism 22, 141–142
metaphors: ecology as 11–12; overview 11
Michael, Linda 57, 59, 61, 64, 69n32
microbial communities 74–75
microbial life 74–75
Mill Race Park 122
Mimus polyglottos 97
mirage 75–76
Molonglo Life 125
Monacella, Rosalea 3
Mondloch, Kate 3
Mondrian, Piet 30–31
Montgomery, Virginia Lee 112
Moss Landing Marine Laboratories 52
Mueck, Ron 68n12
multispecies design 2, 17–20
mutual flourishing 30–31
mutualism 3, 52
mutual relationship 123

Naess, Arne 19
Nagel, Thomas 109
natural habitat 120
"naturalizing" gardens 45
natural/unnatural habitats 120
Nearly Beloved exhibition 60
"Neoprotozoic Oxygenation Event" (NOE) 86, 87
The New York Times 95
"Nocturnal Fugue" 108–109

non-humans 115–126; charisma 118; in commensal community 127; creatures 57; inhabitants 130; nature/culture 145; relationship between humans and 128; sentience 128; spatial landscape 130; species 116, 123; universal rights of 116
North America 21, 96, 129
no-till landscape 131

Oceanix City 52
OctoNation 112
ommatidia 140
The Once and Future World (MacKinnon) 128
open ended space of becoming 132
Orang Laut 42–43
other-than-human species 1; co-creating music with 4
Oxman, Neri 28–29
Oystertecture 47

painting 148
Panda Project 112
parasympathetic nervous system 141
Pasteur, Louis 82
Pentecost, Claire 56
Piccinini, Patricia 3, 56–67
Pickering, Andrew 24, 30–31
Pickett, Steward 2
piñon 150–157; inventory *157*; *ips confusus* 154; juniper-piñon ecosystem 151; need cold 152; need fire 152; science about 153; skeleton *150*; two-needle 151
plants 86–87; evolutions of 87–88; extinctions of 87–88
Plasmid Region exhibition 57, 59, 61–62, 69n26
plover 129–130
pluricellular organisms 86–87
Pollinator Pathmaker project 112
postanthropocentric posthumanism 64–67, 68n6
The Posthuman (Braidotti) 128
The Posthuman (Braidotti) 65, 68n6
post-humanism 128; postanthropocentric 64–67, 68n6
post-humanist philosophy 128–129
post human relations 128
Prakash, Manu 112
primitive hypertext 149
progressive contextualization 11
Protein Lattice-Subset Red, Portrait 66, 70n44
protists 84–85, 86–87
prototyping regulatory change 53–55
Prum, Richard 94, 97
Purple Martin 119

Quaternary Period 73
quintessential anthropocentrism 66
Quirk, Joe 51

racism 12, 64
Rainosek Gallery in Albuquerque 133
Ramankutty, Navin 17
range, defined 95
Rebuild by Design Hurricane Sandy Design Competition 47, 48
reconciliation, transpecies design 20–21
regeneration, transpecies design 20
regenerative agriculture 131
relational becoming 28
restorative practices 141
restoring ecosystems 20
Retallack, Joan 102–103
Rhino models 124
rights of humans 116
Rio Conservation and Sustainability Science Centre (Rio, Brazil) 20
Rising Currents exhibition 47, 48
Roche, François 122
Roeske, Tina 96
Rossi, Aldo 51
Rothenberg, David 4

Saitowitz, Stanley 122
salt formations 71–77; geological time 73–74; microbial life 74–75; mirage 75–76
Santa Fe Art Institute 133
Saraceno, Tomás 110
Sardet, Christian 3
SCAPE 47
Sceloporus lizards 146–147
scent/sound/taste/movement synesthesia 147
Schwegler 115
scientific insights 145
Scientific Reports 141
scientists: and artists relationship 145; and species 145
seasteading 51
Seasteading: How Floating Nations Will Restore the Environment, Enrich the Poor, Cure the Sick, and Liberate Humanity from Politicians (Quirk and Friedman) 51
Seasteading Institute 51
seductive consumerism 63
seed-established landscapes 131
"Semi-Living" 68n14
Seymour, Nicole 65, 70n41
Shelly, Mary 59
Shorter, Wayne 101
showmanship, defined 95
Silk Pavilion 111
Singapore: biodiversity and ecosystem 44; biophilic design 39, 44; "City in Nature" framework 40–42, 45; cohabitation of tidal ecologies 39–48; "Garden City" vision 44–45; *Green Plan 2030* 40; land reclamation 42–44; National Parks (NParks) 40; Orang Laut 42–43; wetlands 46–47
Singer, Peter 18, 25
Siren Mole (SO2) 60, 68n18
slow spaces 132
smooth surface 131–132
Snake Detection Theory 148
speciesism 18, 25
Spider web 110–111
squeeker (mouse coach) 108, *109*
squid map 105–106, *106*
Still Life with Stem Cells 58–59, 61–62
Strand, Sophie 136, 138
Strassburg, Bernardo 20
Survival of the Beautiful (Rothenberg) 93
Suzuki, Wendy 141

Team WAF (Precautions) 57, 62, 68n9
technology 111–112
technosciences 56
temporality 29–30
Tencent 113
Terreform ONE 21, 33, 35, 36
Thacker, Eugene 56
Thian Hock Keng Temple 42
thin line 129–131
Tibetan Buddhism 138
tidal ecologies 39–48
Tifft Nature Preserve 120
Tissue Culture and Art Project (TC&A) 68n14
Todt, Dietmar 100
tone, defined 95
traditional humanism 64
transenvironmentalism 2
transpecies aesthetic 93–103
transpecies co-creation 110–111
transpecies design 2, 4–5, 17–20; and biomimicry 24–31; conceptualizing 24–27; and decentering in ecology 13–14; defined 2; and ecological science 9–14; ethical characteristics 25; ethical dimension of 3; ontological characteristics 25
transpecies dialogues of art 105–113
transpecies ethics 25
transpeciesism 2, 25
transpecies urban theory 24
Triton City 51
Truth and Reconciliation Report 21
Tschumi, Bernard 118
Tuttle, Merlin 112–113
two-needle piñon 151

umwelt 74, 118, 123, 139–140, 145
unbecoming: human 56–67; humanism 63–64
unconditional hospitality 4, 127–128, 130, 132
uneasy nature 64–67
unfolding process 136–138

United Nations Environment Program 43
United Nations Human Settlements Programme (UN-Habitat) 52
United Nations International Panel on Climate Change (IPCC) 90
unnatural habitat 120
urbanization 17; rate of 121
urban landscapes 130

Vacanti mouse 66, *67*, 70n43
Van Valkenburgh, Michael 20, 122
variety, defined 95
vector-borne diseases 1
Vertical Forest (Boeri) 21
visual mythology 140–141
von Uexküll, Jakob Johann 74, 118, 123, 140

Walko, Sarah 4
Wall Street Journal 141
We Are Family exhibition 56–61, 68n7

Wetherwax, Peter 148
wetlands 46–47
When My Baby (When My Baby) 62
Whitelaw, Mitchell 124–125
Whyte, David 141
Wigglesworth, Sarah 115
wilderness protections 130
Willow, Gabriel 4
Wolch, Jennifer 2
Wolfe, Carey 127–128, 130, 131
World Bank 17
"worlding" 71
"world-making" 72

The Young Family 58, 61, 68n3

Zaretsky, Adrian Parr 2, 4, 90–92, 144–149
Zaretsky, Michael 1–5
zoocentrism 25
zoonotic diseases 1
Zurr, Ionat 68n14